Short Courses

 Microorganisms, Fungi, and Plants
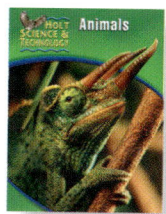 Animals
 Cells, Heredity, and Classification
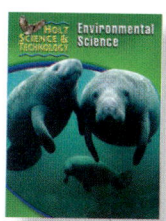 Environmental Science
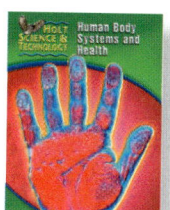 Human Body Systems and Health

 Inside the Restless Earth
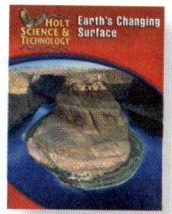 Earth's Changing Surface
 Water on Earth
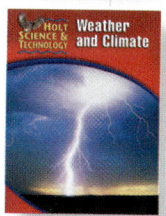 Weather and Climate
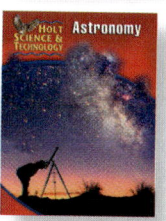 Astronomy

 Introduction to Matter
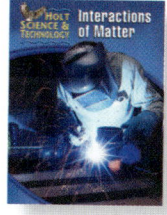 Interactions of Matter
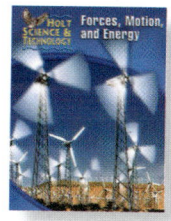 Forces, Motion, and Energy
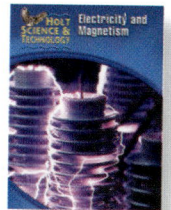 Electricity and Magnetism
 Sound and Light

Teacher Edition WALK-THROUGH

Student Edition CONTENTS IN BRIEF

HOLT, RINEHART AND WINSTON

A Harcourt Education Company

Orlando • **Austin** • New York • San Diego • Toronto • London

Designed to meet the needs of all students

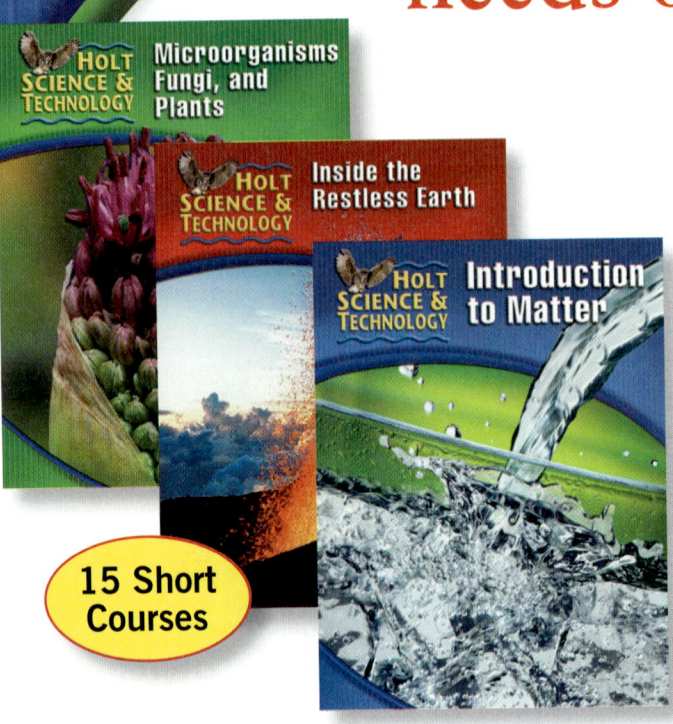

15 Short Courses

Holt Science & Technology: Short Course Series allows you to match your curriculum by choosing from 15 books covering life, earth, and physical sciences. The program reflects current curriculum developments and includes the strongest skills-development strand of any middle school science series. Students of all abilities will develop skills that they can use both in science as well as in other courses.

STUDENTS OF ALL ABILITIES RECEIVE THE READING HELP AND TAILORED INSTRUCTION THEY NEED.

- The *Student Edition* is accessible with a clean, easy-to-follow design and highlighted vocabulary words.
- Inclusion strategies and different learning styles help support all learners.
- Comprehensive **Section** and **Chapter Reviews** and **Standardized Test Preparation** allow students to practice their test-taking skills.
- **Reading Comprehension Guide** and **Guided Reading Audio CDs** help students better understand the content.

CROSS-DISCIPLINARY CONNECTIONS LET STUDENTS SEE HOW SCIENCE RELATES TO OTHER DISCIPLINES.

- **Mathematics, reading,** and **writing skills** are integrated throughout the program.
- Cross-discipline **Connection To** features show students how science relates to language arts, social studies, and other sciences.

A FLEXIBLE LABORATORY PROGRAM HELPS STUDENTS BUILD IMPORTANT INQUIRY AND CRITICAL-THINKING SKILLS.

- The laboratory program includes labs in each chapter, labs in the **LabBook** at the end of the text, six different lab books, and **Video Labs.**
- All labs are teacher-tested and rated by difficulty in the *Teacher Edition,* so you can be sure the labs will be appropriate for your students.
- A variety of labs, from **Inquiry Labs** to **Skills Practice Labs,** helps you meet the needs of your curriculum and work within the time constraints of your teaching schedule.

INTEGRATED TECHNOLOGY AND ONLINE RESOURCES EXPAND LEARNING BEYOND CLASSROOM WALLS.

- An **Enhanced Online Edition** or **CD-ROM Version** of the student text lightens your students' load.

- **SciLinks,** a Web service developed and maintained by the National Science Teachers Association (NSTA), contains current prescreened links directly related to the textbook.

- **Brain Food Video Quizzes** on videotape and DVD are game-show style quizzes that assess students' progress and motivate them to study.

- The **One-stop Planner® CD-ROM** with **Exam View® Test Generator** contains all of the resources you need including an *Interactive Teacher Edition,* worksheets, customizable lesson plans, **Holt Calendar Planner,** a powerful test generator, **Lab Materials QuickList Software,** and more.

- Spanish Resources include **Guided Reading Audio CD** in Spanish.

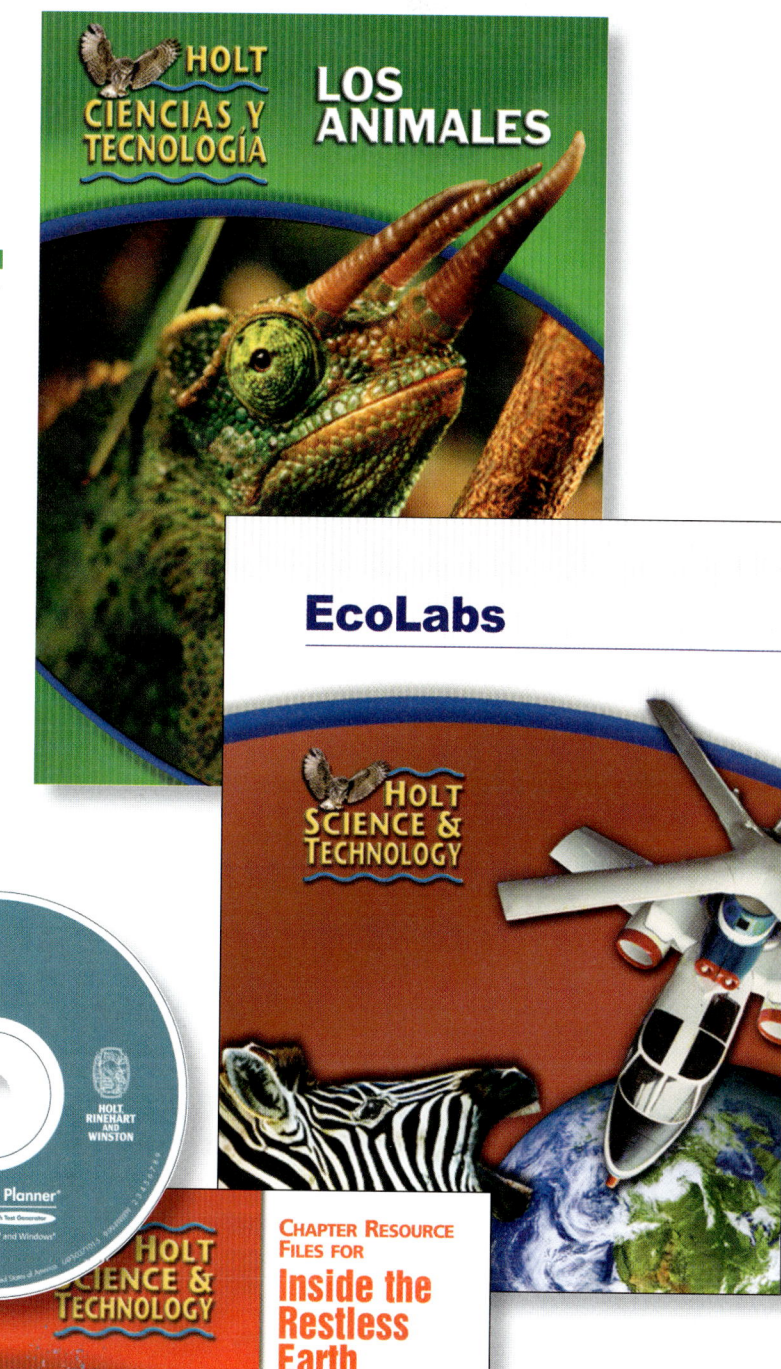

HOLT **CIENCIAS Y TECNOLOGÍA** — LOS ANIMALES

EcoLabs — HOLT SCIENCE & TECHNOLOGY

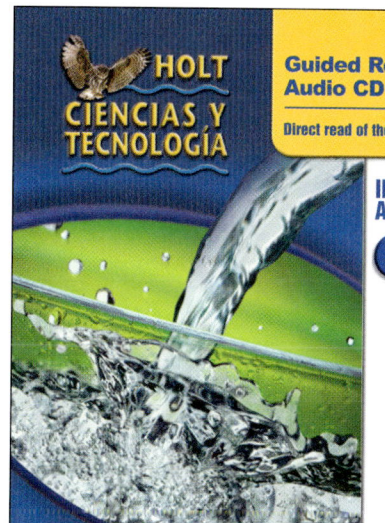

HOLT **CIENCIAS Y TECNOLOGÍA**

Guided Reading Audio CD Program

Direct read of the student text

INTRODUCCIÓN A LA MATERIA

K

HOLT SCIENCE & TECHNOLOGY

CHAPTER RESOURCE FILES FOR

Inside the Restless Earth

Skills Worksheets
- Directed Reading A
- Directed Reading B
- Vocabulary & Notes
- Section Reviews
- Chapter Review
- Reinforcement
- Critical Thinking

Assessments
- Section Quizzes
- Chapter Test A
- Chapter Test B
- Chapter Test C
- Performance-Based Assessment
- Standardized Test Preparation

Labs and Activities
- Datasheets for In-Text Labs
- Datasheets for Quick Labs
- Datasheets for LabBook
- Vocabulary Activity
- SciLinks® Activity

Teacher Resources
- Teacher Notes for Performance-Based Assessment
- Lab Notes and Answers
- Answer Keys
- Lesson Plans
- Test Item Listing for ExamView® Test Generator
- Teaching Transparencies
- Chapter Starter Transparencies
- Bellringer Transparencies
- Concept Mapping Transparencies

Life Science

 A MICROORGANISMS, FUNGI, AND PLANTS

 B ANIMALS

PROGRAM SCOPE AND SEQUENCE

Selecting the right books for your course is easy. Just review the topics presented in each book to determine the best match to your district curriculum.

C CELLS, HEREDITY, & CLASSIFICATION

Cells: The Basic Units of Life
- Cells, tissues, and organs
- Cell theory
- Surface-to-volume ratio
- Prokaryotic versus eukaryotic cells
- Cell organelles

The Cell in Action
- Diffusion and osmosis
- Passive versus active transport
- Endocytosis versus exocytosis
- Photosynthesis
- Cellular respiration and fermentation
- Cell cycle

Heredity
- Dominant versus recessive traits
- Genes and alleles
- Genotype, phenotype, the Punnett square and probability
- Meiosis
- Determination of sex

Genes and Gene Technology
- Structure of DNA
- Protein synthesis
- Mutations
- Heredity disorders and genetic counseling

The Evolution of Living Things
- Adaptations and species
- Evidence for evolution
- Darwin's work and natural selection
- Formation of new species

The History of Life on Earth
- Geologic time scale and extinctions
- Plate tectonics
- Human evolution

Classification
- Levels of classification
- Cladistic diagrams
- Dichotomous keys
- Characteristics of the six kingdoms

D HUMAN BODY SYSTEMS & HEALTH

Body Organization and Structure
- Homeostasis
- Types of tissue
- Organ systems
- Structure and function of the skeletal system, muscular system, and integumentary system

Circulation and Respiration
- Structure and function of the cardiovascular system, lymphatic system, and respiratory system
- Respiratory disorders

The Digestive and Urinary Systems
- Structure and function of the digestive system
- Structure and function of the urinary system

Communication and Control
- Structure and function of the nervous system and endocrine system
- The senses
- Structure and function of the eye and ear

Reproduction and Development
- Asexual versus sexual reproduction
- Internal versus external fertilization
- Structure and function of the human male and female reproductive systems
- Fertilization, placental development, and embryo growth
- Stages of human life

Body Defenses and Disease
- Types of diseases
- Vaccines and immunity
- Structure and function of the immune system
- Autoimmune diseases, cancer, and AIDS

Staying Healthy
- Nutrition and reading food labels
- Alcohol and drug effects on the body
- Hygiene, exercise, and first aid

E ENVIRONMENTAL SCIENCE

Interactions of Living Things
- Biotic versus abiotic parts of the environment
- Producers, consumers, and decomposers
- Food chains and food webs
- Factors limiting population growth
- Predator-prey relationships
- Symbiosis and coevolution

Cycles in Nature
- Water cycle
- Carbon cycle
- Nitrogen cycle
- Ecological succession

The Earth's Ecosystems
- Kinds of land and water biomes
- Marine ecosystems
- Freshwater ecosystems

Environmental Problems and Solutions
- Types of pollutants
- Types of resources
- Conservation practices
- Species protection

Energy Resources
- Types of resources
- Energy resources and pollution
- Alternative energy resources

Earth Science

H WATER ON EARTH

 I WEATHER AND CLIMATE

J ASTRONOMY

The Flow of Fresh Water
- Water cycle
- River systems
- Stream erosion
- Life cycle of rivers
- Deposition
- Aquifers, springs, and wells
- Ground water
- Water treatment and pollution

The Atmosphere
- Structure of the atmosphere
- Air pressure
- Radiation, convection, and conduction
- Greenhouse effect and global warming
- Characteristics of winds
- Types of winds
- Air pollution

Studying Space
- Astronomy
- Keeping time
- Types of telescope
- Radioastronomy
- Mapping the stars
- Scales of the universe

Exploring the Oceans
- Properties and characteristics of the oceans
- Features of the ocean floor
- Ocean ecology
- Ocean resources and pollution

Understanding Weather
- Water cycle
- Humidity
- Types of clouds
- Types of precipitation
- Air masses and fronts
- Storms, tornadoes, and hurricanes
- Weather forecasting
- Weather maps

Stars, Galaxies, and the Universe
- Composition of stars
- Classification of stars
- Star brightness, distance, and motions
- H-R diagram
- Life cycle of stars
- Types of galaxies
- Theories on the formation of the universe

The Movement of Ocean Water
- Types of currents
- Characteristics of waves
- Types of ocean waves
- Tides

Climate
- Weather versus climate
- Seasons and latitude
- Prevailing winds
- Earth's biomes
- Earth's climate zones
- Ice ages
- Global warming
- Greenhouse effect

Formation of the Solar System
- Birth of the solar system
- Structure of the sun
- Fusion
- Earth's structure and atmosphere
- Planetary motion
- Newton's Law of Universal Gravitation

A Family of Planets
- Properties and characteristics of the planets
- Properties and characteristics of moons
- Comets, asteroids, and meteoroids

Exploring Space
- Rocketry and artificial satellites
- Types of Earth orbit
- Space probes and space exploration

Physical Science

FORCES, MOTION, AND ENERGY

Matter in Motion
- Speed, velocity, and acceleration
- Measuring force
- Friction
- Mass versus weight

Forces in Motion
- Terminal velocity and free fall
- Projectile motion
- Inertia
- Momentum

Forces in Fluids
- Properties in fluids
- Atmospheric pressure
- Density
- Pascal's principle
- Buoyant force
- Archimedes' principle
- Bernoulli's principle

Work and Machines
- Measuring work
- Measuring power
- Types of machines
- Mechanical advantage
- Mechanical efficiency

Energy and Energy Resources
- Forms of energy
- Energy conversions
- Law of conservation of energy
- Energy resources

Heat and Heat Technology
- Heat versus temperature
- Thermal expansion
- Absolute zero
- Conduction, convection, radiation
- Conductors versus insulators
- Specific heat capacity
- Changes of state
- Heat engines
- Thermal pollution

N **ELECTRICITY AND MAGNETISM**

Introduction to Electricity
- Law of electric charges
- Conduction versus induction
- Static electricity
- Potential difference
- Cells, batteries, and photocells
- Thermocouples
- Voltage, current, and resistance
- Electric power
- Types of circuits

Electromagnetism
- Properties of magnets
- Magnetic force
- Electromagnetism
- Solenoids and electric motors
- Electromagnetic induction
- Generators and transformers

Electronic Technology
- Properties of semiconductors
- Integrated circuits
- Diodes and transistors
- Analog versus digital signals
- Microprocessors
- Features of computers

O **SOUND AND LIGHT**

The Energy of Waves
- Properties of waves
- Types of waves
- Reflection and refraction
- Diffraction and interference
- Standing waves and resonance

The Nature of Sound
- Properties of sound waves
- Structure of the human ear
- Pitch and the Doppler effect
- Infrasonic versus ultrasonic sound
- Sound reflection and echolocation
- Sound barrier
- Interference, resonance, diffraction, and standing waves
- Sound quality of instruments

The Nature of Light
- Electromagnetic waves
- Electromagnetic spectrum
- Law of reflection
- Absorption and scattering
- Reflection and refraction
- Diffraction and interference

Light and Our World
- Luminosity
- Types of lighting
- Types of mirrors and lenses
- Focal point
- Structure of the human eye
- Lasers and holograms

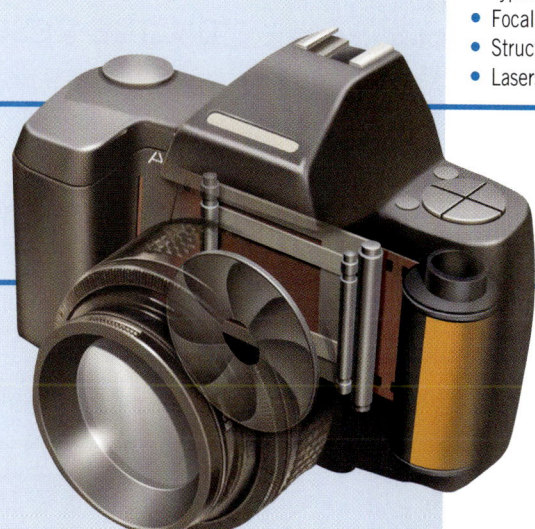

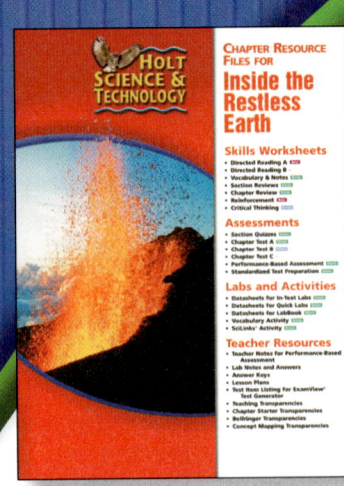

Program resources make teaching and learning easier.

CHAPTER RESOURCES

A *Chapter Resources book* accompanies each of the 15 *Short Courses*. Here you'll find everything you need to make sure your students are getting the most out of learning science—all in one book.

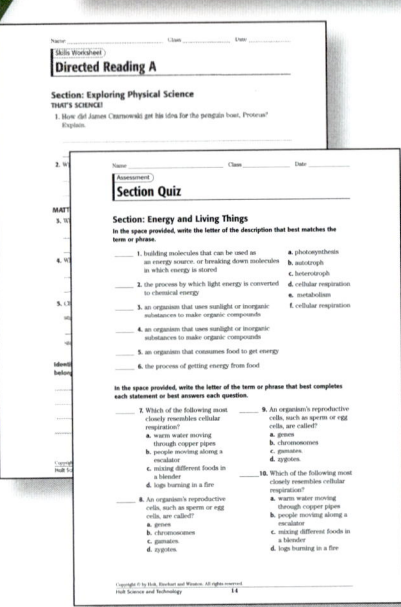

Skills Worksheets
- Directed Reading A: Basic
- Directed Reading B: Special Needs
- Vocabulary and Chapter Summary
- Section Reviews
- Chapter Reviews
- Reinforcement
- Critical Thinking

Labs & Activities
- Datasheets for Chapter Labs
- Datasheets for Quick Labs
- Datasheets for LabBook
- Vocabulary Activity
- SciLinks® Activity

Assessments
- Section Quizzes
- Chapter Tests A: General
- Chapter Tests B: Advanced
- Chapter Tests C: Special Needs
- Performance-Based Assessments
- Standardized Test Preparation

Teacher Resources
- Lab Notes and Answers
- Teacher Notes for Performance-Based Assessment
- Answer Keys
- Lesson Plans
- Test Item Listing for ExamView® Test Generator
- Full-color **Teaching Transparencies,** plus section **Bellringers, Concept Mapping,** and **Chapter Starter Transparencies.**

SPANISH RESOURCES

Spanish materials are available for each *Short Course:*

- *Student Edition*
- **Spanish Resources** booklet contains worksheets and assessments translated into Spanish with an English **Answer Key.**
- **Guided Reading Audio CD Program**

ONLINE RESOURCES

- *Enhanced Online Editions* engage students and assist teachers with a host of interactive features that are available anytime and anywhere you can connect to the Internet.
- **CNNStudentNews.com** provides award-winning news and information for both teachers and students.
- **SciLinks**—a Web service developed and maintained by the National Science Teachers Association—links you and your students to up-to-date online resources directly related to chapter topics.
- **go.hrw.com** links you and your students to online chapter activities and resources.
- **Current Science** articles relate to students' lives.

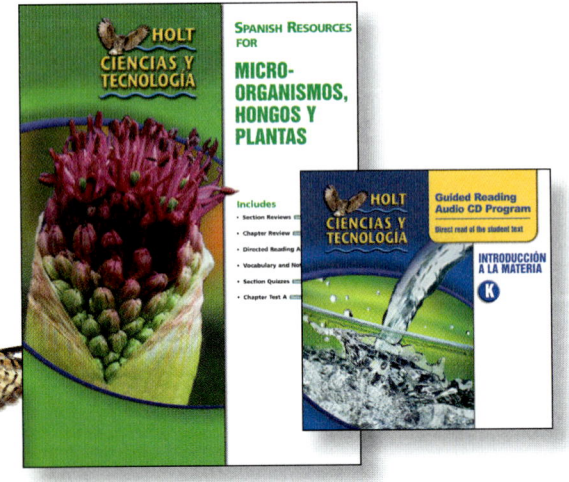

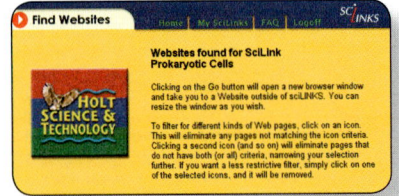

T10

ADDITIONAL LAB AND SKILLS RESOURCES

- *Calculator-Based Labs* incorporates scientific instruments, offering students insight into modern scientific investigation.
- *EcoLabs & Field Activities* develops awareness of the natural world.
- *Holt Science Skills Workshop: Reading in the Content Area* contains exercises that target reading skills key.
- *Inquiry Labs* taps students' natural curiosity and creativity with a focus on the process of discovery.
- *Labs You Can Eat* safely incorporates edible items into the classroom.
- *Long-Term Projects & Research Ideas* extends and enriches lessons.
- *Math Skills for Science* provides additional explanations, examples, and math problems so students can develop their skills.
- *Science Skills Worksheets* helps your students hone important learning skills.
- *Whiz-Bang Demonstrations* gets your students' attention at the beginning of a lesson.

ADDITIONAL RESOURCES

- *Assessment Checklists & Rubrics* gives you guidelines for evaluating students' progress.
- *Holt Anthology of Science Fiction* sparks your students' imaginations with thought-provoking stories.
- *Holt Science Posters* visually reinforces scientific concepts and themes with seven colorful posters including **The Periodic Table of the Elements.**

- *Professional Reference for Teachers* contains professional articles that discuss a variety of topics, such as classroom management.
- *Program Introduction Resource File* explains the program and its features and provides several additional references, including lab safety, scoring rubrics, and more.
- *Science Fair Guide* gives teachers, students, and parents tips for planning and assisting in a science fair.
- *Science Puzzlers, Twisters & Teasers* activities challenge students to think about science concepts in different ways.

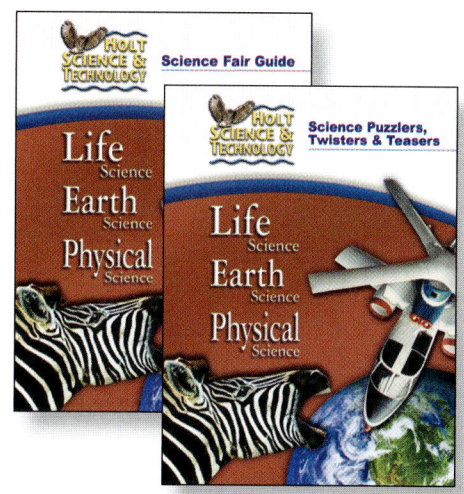

TECHNOLOGY RESOURCES

- *CNN Presents Science in the News: Video Library* helps students see the impact of science on their everyday lives with actual news video clips.
  - Multicultural Connections
  - Science, Technology & Society
  - Scientists in Action
  - Eye on the Environment
- *Guided Reading Audio CD Program*, available in English and Spanish, provides students with a direct read of each section.
- *HRW Earth Science Videotape* takes your students on a geology "field trip" with full-motion video.
- *Interactive Explorations CD-ROM Program* develops students' inquiry and decision-making skills as they investigate science phenomena in a virtual lab setting.

- *One-Stop Planner CD-ROM*® organizes everything you need on one disc, including printable worksheets, customizable lesson plans, a powerful test generator, **PowerPoint**® **Resources, Lab Materials QuickList Software, Holt Calendar Planner, Interactive Teacher Edition,** and more.
- *Science Tutor CD-ROMs* help students practice what they learn with immediate feedback.
- *Lab Videos* make it easier to integrate more experiments into your lessons without the preparation time and costs. Available on DVD and VHS.
- **Brain Food Video Quizzes** are game-show style quizzes that assess students' progress. Available on DVD and VHS.
- *Visual Concepts CD-ROMs* include graphics, animations, and movie clips that demonstrate key chapter concepts.

Science and Math Worksheets

Science Skills Worksheets: Thinking Skills

BEING FLEXIBLE
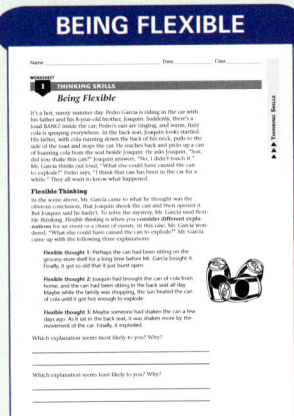

USING YOUR SENSES
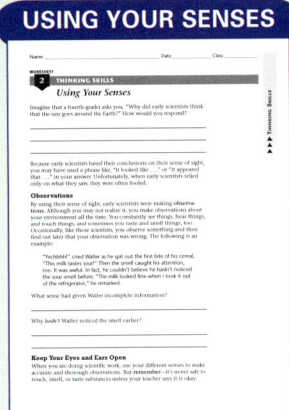

THINKING OBJECTIVELY
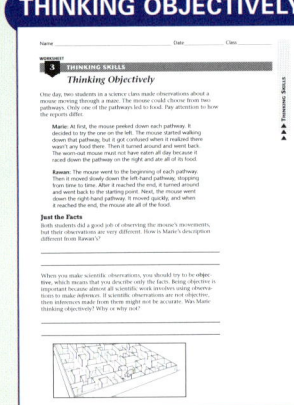

UNDERSTANDING BIAS
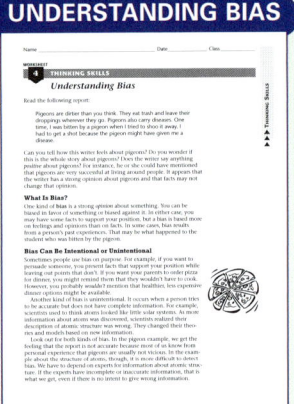

USING LOGIC
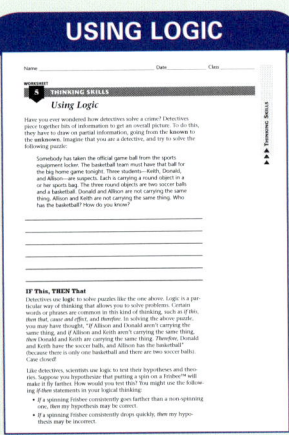

BOOSTING YOUR MEMORY
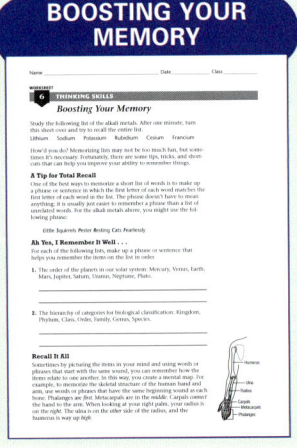

IMPROVING YOUR STUDY HABITS

READING A SCIENCE TEXTBOOK
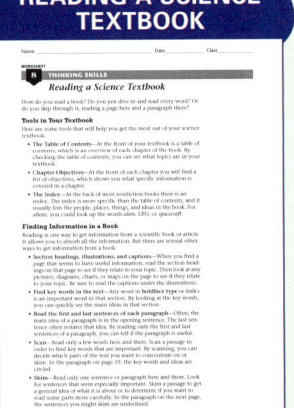

Science Skills Worksheets: Experimenting Skills

SAFETY RULES!

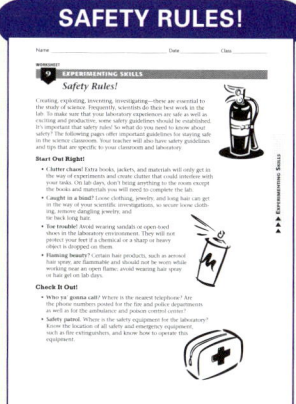

DOING A LAB WRITE-UP

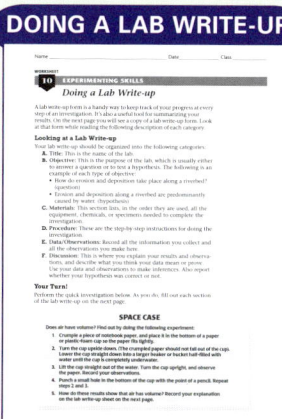

UNDERSTANDING VARIABLES

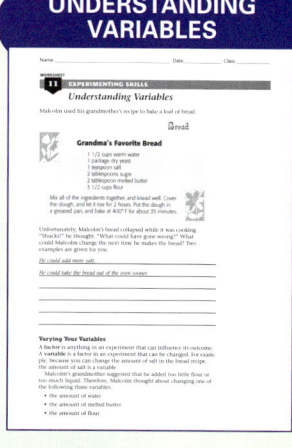

WORKING WITH HYPOTHESES

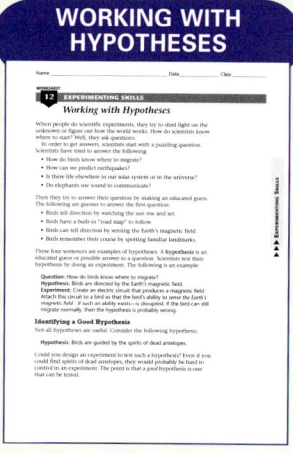

DESIGNING AN EXPERIMENT

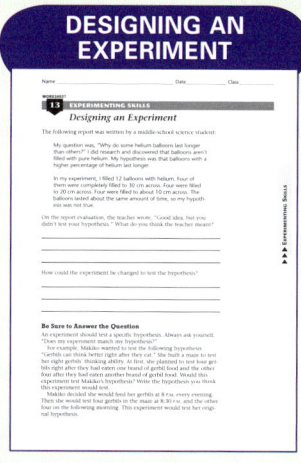

USING THE INTERNATIONAL SYSTEM OF UNITS (SI)

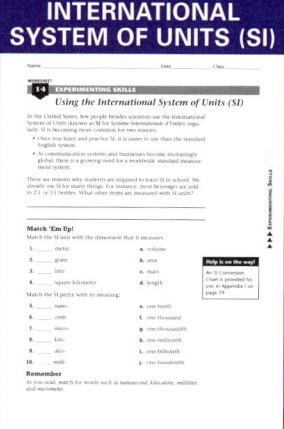

MEASURING

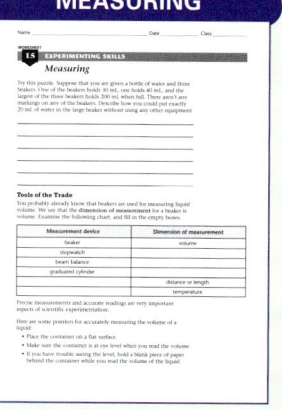

Science Skills Worksheets: Researching Skills

CHOOSING YOUR TOPIC

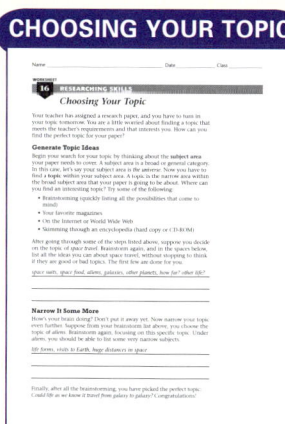

ORGANIZING YOUR RESEARCH

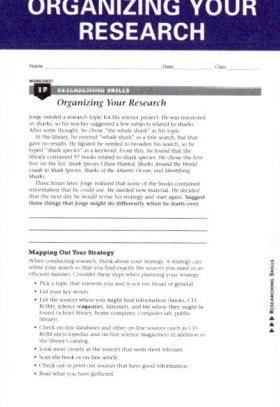

FINDING USEFUL SOURCES

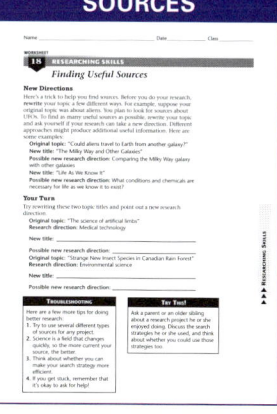

RESEARCHING ON THE WEB

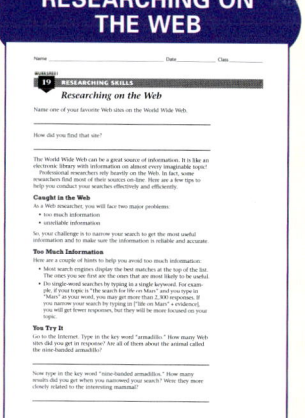

Science Skills Worksheets: Researching Skills (continued)

IDENTIFYING BIAS

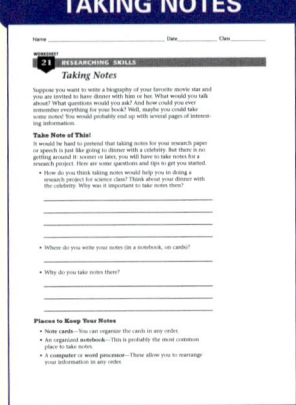

TAKING NOTES

Science Skills Worksheets: Communicating Skills

SCIENCE WRITING

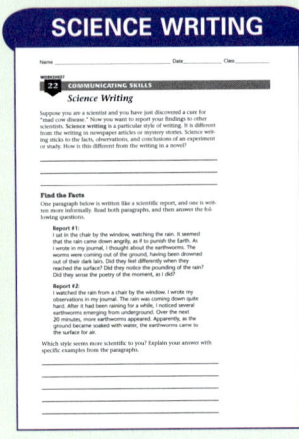

SCIENCE DRAWING

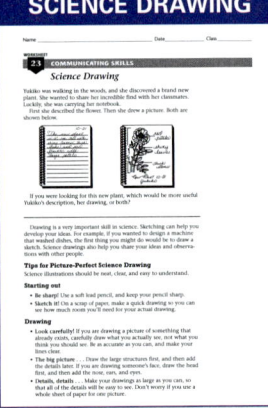

USING MODELS TO COMMUNICATE

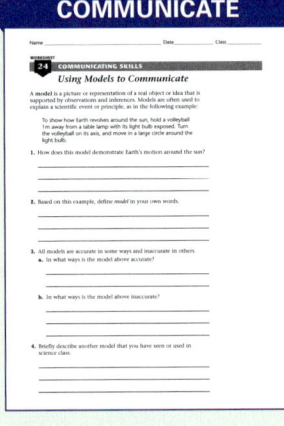

INTRODUCTION TO GRAPHS

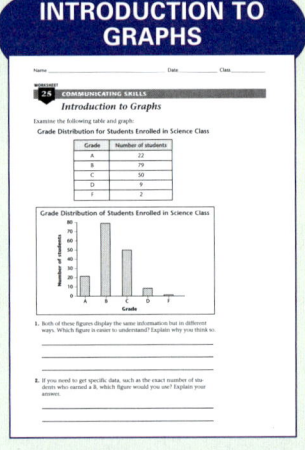

GRASPING GRAPHING

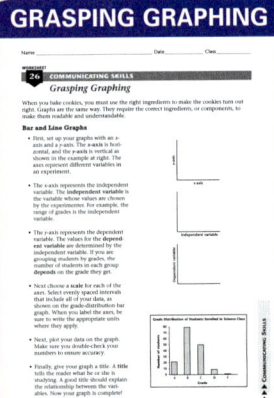

INTERPRETING YOUR DATA

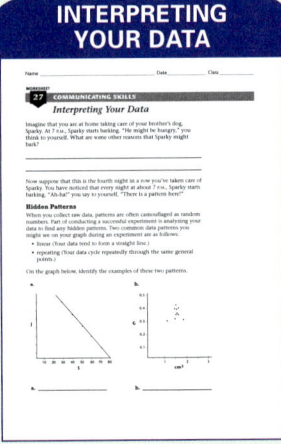

RECOGNIZING BIAS IN GRAPHS

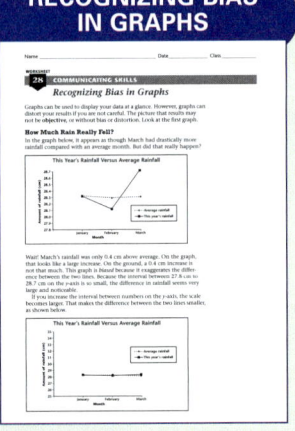

MAKING DATA MEANINGFUL

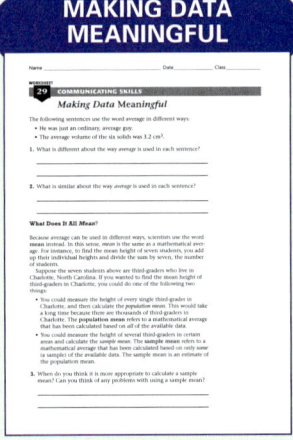

HINTS FOR ORAL PRESENTATIONS

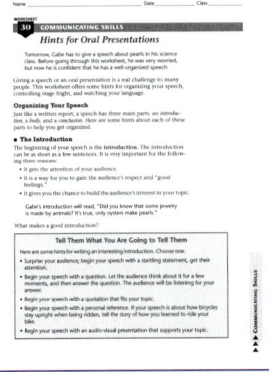

Math Skills for Science

ADDITION AND SUBTRACTION

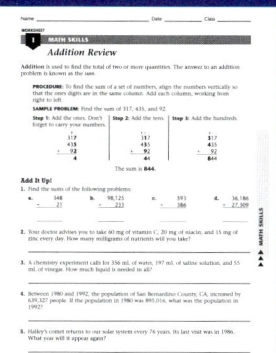

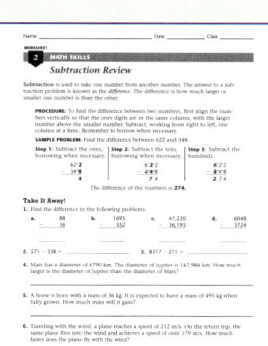

MULTIPLICATION

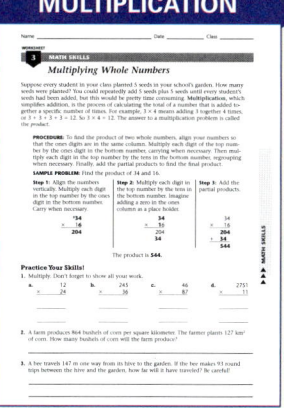

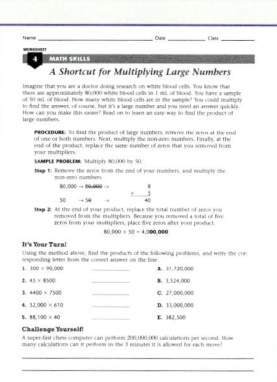

DIVISION

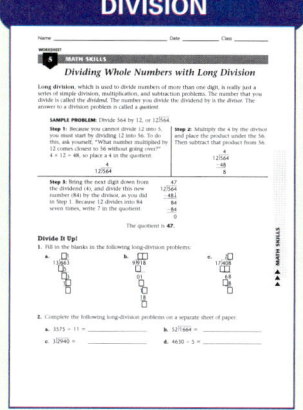

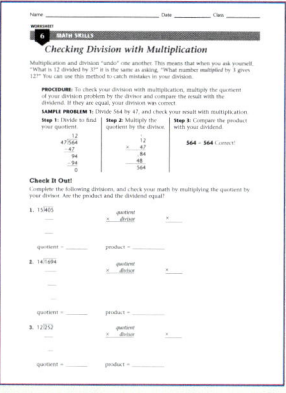

AVERAGES

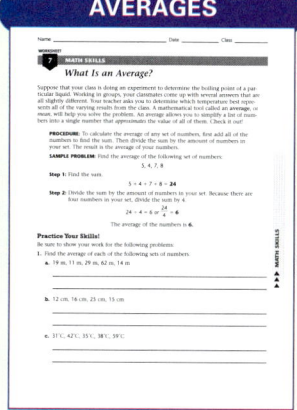

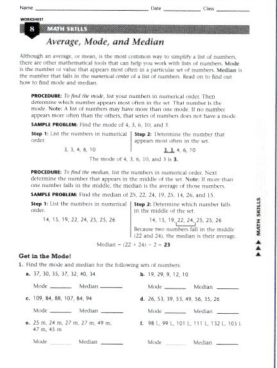

POSITIVE AND NEGATIVE NUMBERS

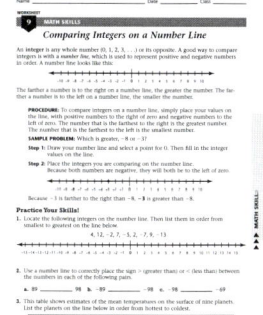

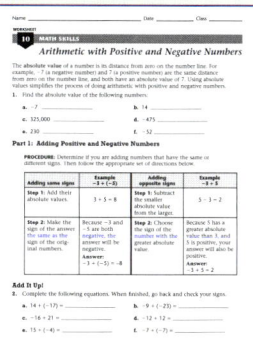

FRACTIONS

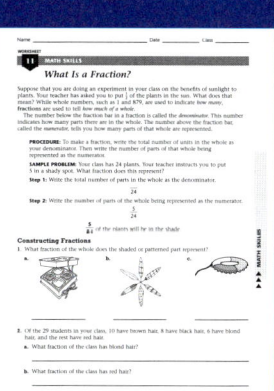

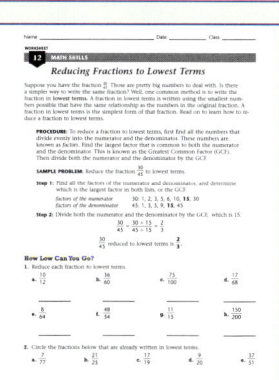

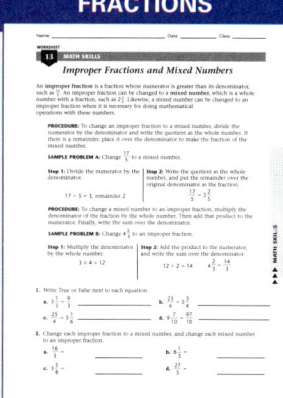

Math Skills for Science (continued)

RATIOS AND PROPORTIONS

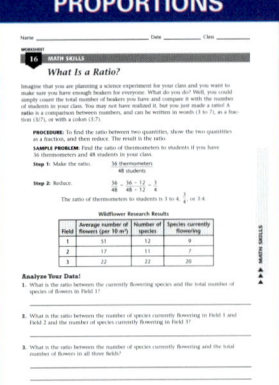

DECIMALS

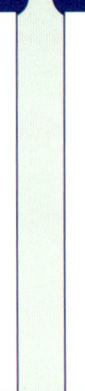

PERCENTAGES

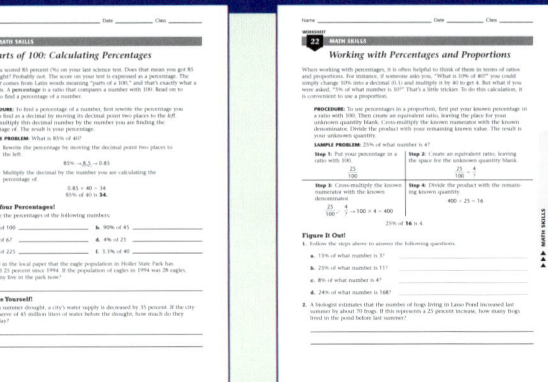

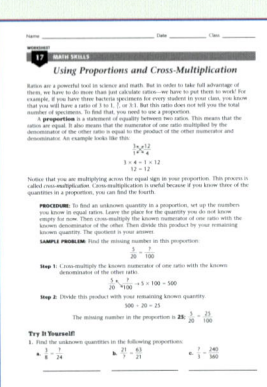

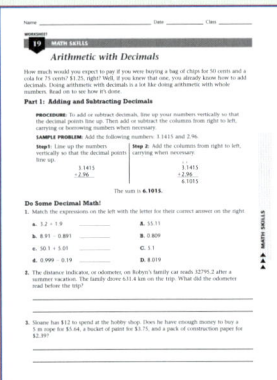

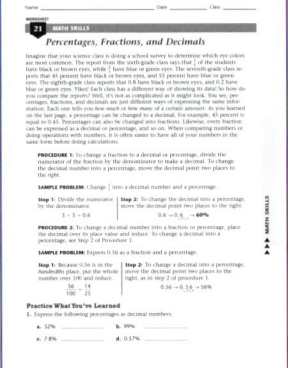

POWERS OF 10

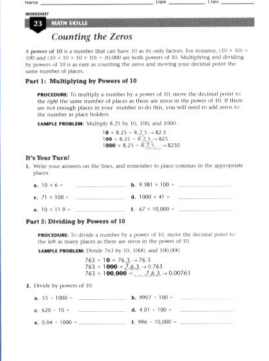

SCIENTIFIC NOTATION

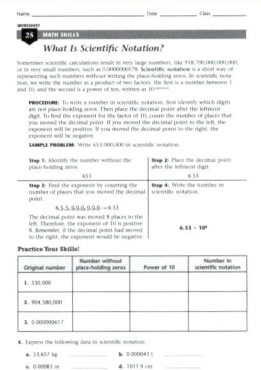

SI MEASUREMENT AND CONVERSION

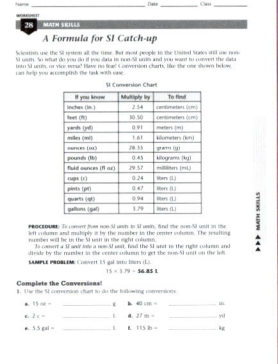

Math Skills for Science (continued)

GEOMETRY

THE UNIT FACTOR AND DIMENSIONAL ANALYSIS

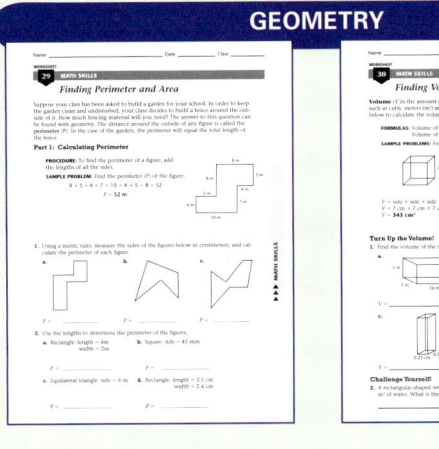

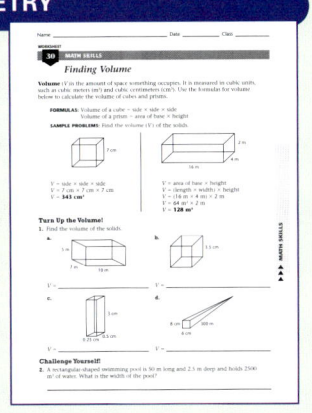

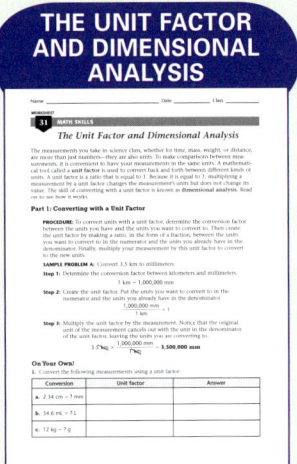

MATH IN SCIENCE: INTEGRATED SCIENCE

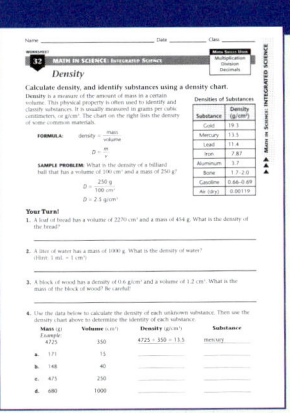

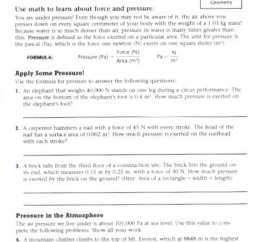

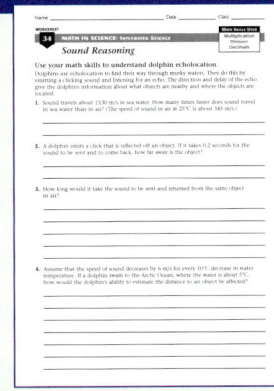

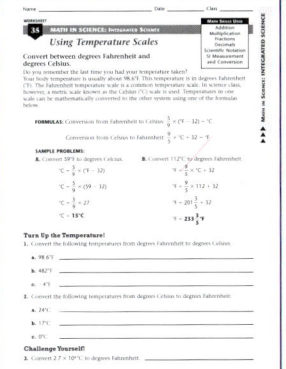

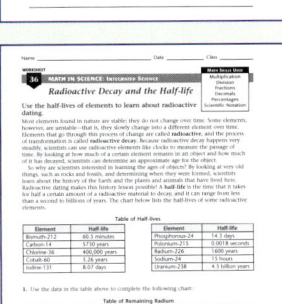

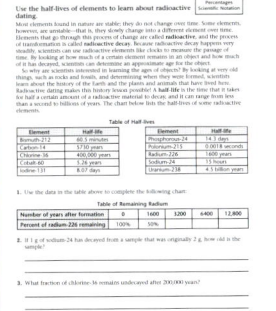

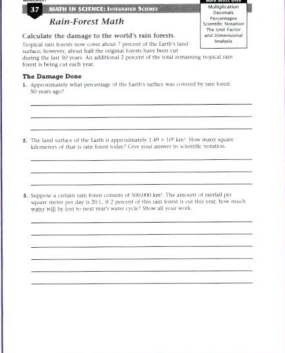

T17

Math Skills for Science (continued)

MATH IN SCIENCE: LIFE SCIENCE

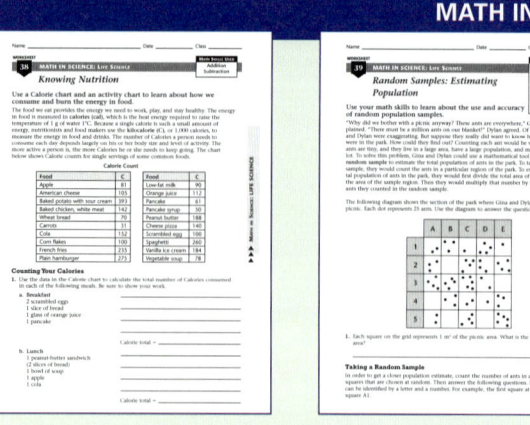

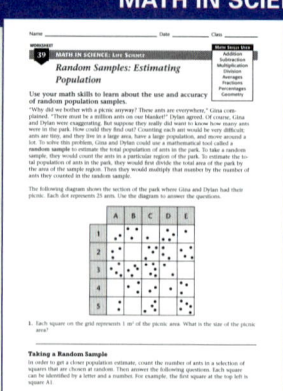

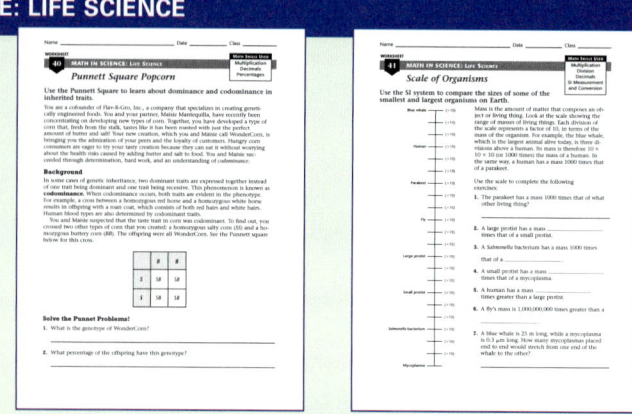

MATH IN SCIENCE: EARTH SCIENCE

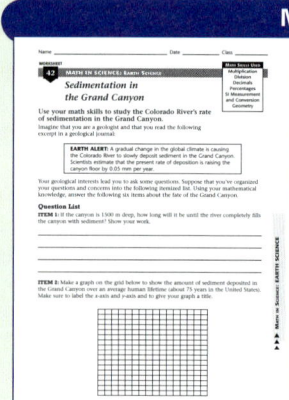

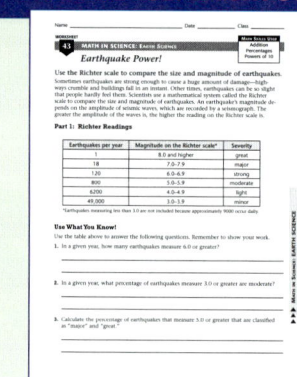

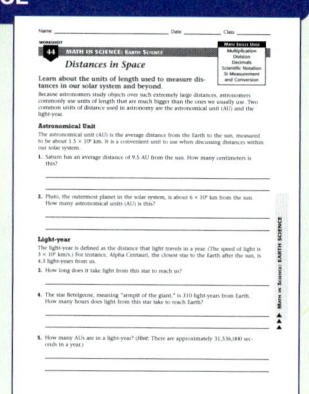

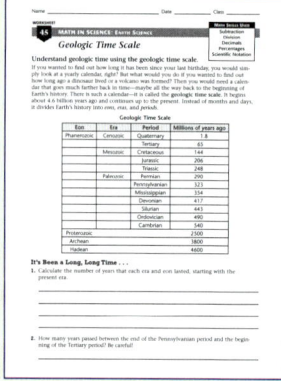

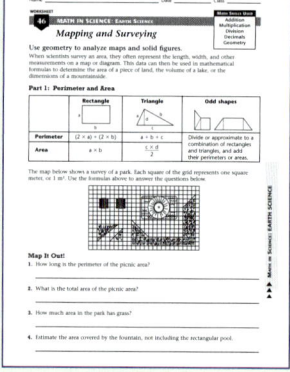

Math Skills for Science (continued)

MATH IN SCIENCE: PHYSICAL SCIENCE

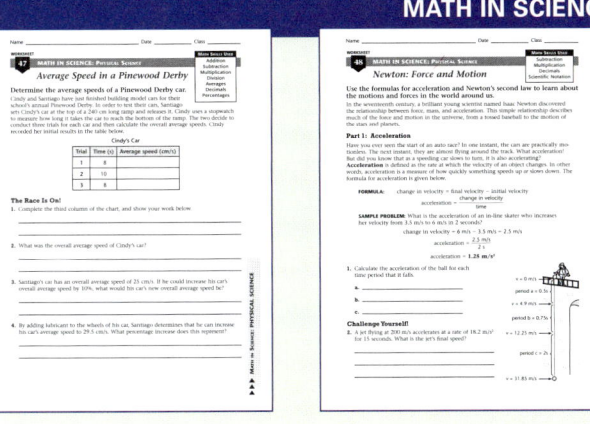

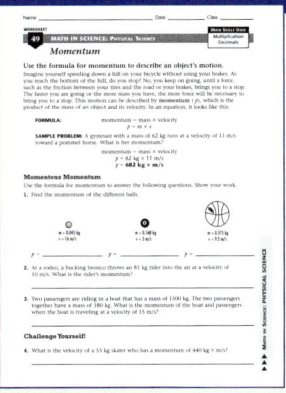

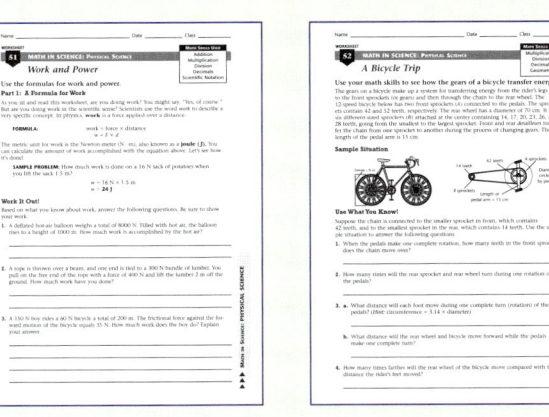

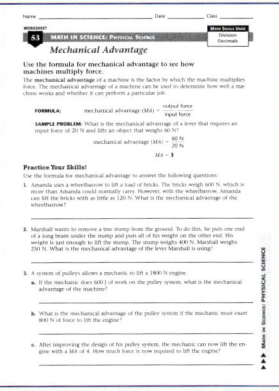

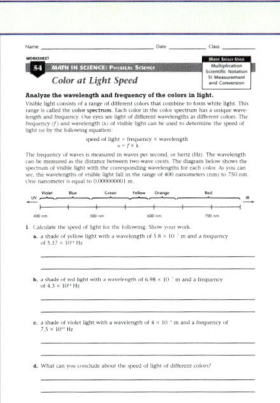

Assessment Checklist & Rubrics

The following is just a sample of over 50 checklists and rubrics contained in this booklet.

RUBRICS FOR WRITTEN WORK

RUBRIC FOR EXPERIMENTS

TEACHER EVALUATION OF COOPERATIVE LEARNING

TEACHER EVALUATION OF STUDENT PROGRESS

National Science Education Standards

The following lists show the chapter correlation of *Holt Science & Technology: Earth's Changing Surface* with the *National Science Education Standards* (grades 5–8).

Unifying Concepts and Processes

Standard	Chapter Correlation	
Evidence, models, and explanation Code: UCP 2	Chapter 1	1.1, 1.2, 1.3
	Chapter 2	2.1
	Chapter 3	3.2, 3.3
Change, constancy, and measurement Code: UCP 3	Chapter 1	1.1, 1.2, 1.3
	Chapter 3	3.1, 3.2, 3.3
Form and function Code: UCP 5	Chapter 1	1.1, 1.2, 1.3

Science as Inquiry

Standard	Chapter Correlation	
Abilities necessary to do scientific inquiry Code: SAI 1	Chapter 1	1.1, 1.2, 1.3
	Chapter 2	2.1, 2.2, 2.3, 2.4
	Chapter 3	3.1, 3.2, 3.3, 3.4
Understandings about scientific inquiry Code: SAI 2	Chapter 2	2.1, 2.2, 2.3

Science and Technology

Standard	Chapter Correlation	
Abilities of technological design Code: ST 1	Chapter 1	1.1, 1.3
Understandings about science and technology Code: ST 2	Chapter 1	1.1, 1.2, 1.3
	Chapter 2	2.4
	Chapter 3	3.2

Science in Personal Perspectives

Standard	Chapter Correlation	
Populations, resources, and environments Code: SPSP 2	Chapter 1	1.3
	Chapter 2	2.4
	Chapter 3	3.1, 3.2
Natural hazards Code: SPSP 3	Chapter 2	2.4
	Chapter 3	3.1, 3.3, 3.4
Risks and benefits Code: SPSP 4	Chapter 2	2.3, 2.4
	Chapter 3	3.4
Science and technology in society Code: SPSP 5	Chapter 1	1.1, 1.2, 1.3
	Chapter 2	2.4

History and Nature of Science

Standard	Chapter Correlation	
Science as a human endeavor Code: HNS 1	Chapter 1	1.1, 1.3
	Chapter 2	2.4
	Chapter 3	3.2
History of science Code: HNS 3	Chapter 1	1.1, 1.2

Earth Science Content Standards

Structure of the Earth System

Standard	Chapter Correlation	
Land forms are the result of a combination of constructive and destructive forces. Constructive forces include crustal deformation, volcanic eruption, and deposition of sediment, while destructive forces include weathering and erosion. Code: ES 1c	**Chapter 2** **Chapter 3**	2.1, 2.2, 2.3 3.1, 3.2, 3.3, 3.4
Some changes in the solid earth can be described as the "rock cycle." Old rocks at the earth's surface weather, forming sediments that are buried, then compacted, heated, and often recrystallized into new rock. Eventually, those new rocks may be brought to the earth's surface by the forces that drive plate motions, and the rock cycle continues. Code: ES 1d	**Chapter 2**	2.1, 2.2
Soil consists of weathered rocks and decomposed organic material from dead plants, animals, and bacteria. Soils are often found in layers, with each having a different chemical composition and texture. Code: ES 1e	**Chapter 2**	2.3
Water is a solvent. As it passes through the water cycle it dissolves minerals and gases and carries them to the oceans. Code: ES 1g	**Chapter 2**	2.3
Living organisms have played many roles in the earth system, including affecting the composition of the atmosphere, producing some types of rocks, and contributing to the weathering of rocks. Code: ES 1k	**Chapter 2**	2.1, 2.3

Earth's History

Standard	Chapter Correlation	
The earth processes we see today, including erosion, movement of lithospheric plates, and changes in atmospheric composition, are similar to those that occurred in the past. Earth history is also influenced by occasional catastrophes, such as the impact of an asteroid or comet. Code: ES 2a	**Chapter 3**	3.2, 3.3

HOLT SCIENCE & TECHNOLOGY

Earth's Changing Surface

HOLT, RINEHART AND WINSTON

A Harcourt Education Company

Orlando • **Austin** • New York • San Diego • Toronto • London

Acknowledgments

Contributing Authors

Kathleen Meehan Berry
Science Chairman
Canon-McMillan School
    District
Canonsburg,
    Pennsylvania

Robert H. Fronk, Ph.D.
Professor
Science and Mathematics
    Education Department
Florida Institute of
    Technology
Melbourne, Florida

**Robert J. Sager, M.S.,
    J.D., L.G.**
*Coordinator and Professor
    of Earth Science*
Pierce College
Lakewood, Washington

Inclusion Specialist

Karen Clay
*Inclusion Specialist
    Consultant*
Boston, Massachusetts

Safety Reviewer

Jack Gerlovich, Ph.D.
Associate Professor
School of Education
Drake University
Des Moines, Iowa

Academic Reviewers

John Brockhaus, Ph.D.
*Professor of Geospatial
    Information Science and
    Director of Geospatial
    Information Science
    Program*
Department of
    Geography and
    Environmental
    Engineering
United States Military
    Academy
West Point, New York

**Steven A. Jennings,
    Ph.D.**
Associate Professor
Geography and
    Environmental Studies
University of Colorado at
    Colorado Springs
Colorado Springs,
    Colorado

**Madeline Micceri
    Mignone, Ph.D.**
Assistant Professor
Natural Science
Dominican College
Orangeburg, New York

Daniel Z. Sui, Ph.D.
Professor
Department of
    Geography
Texas A&M University
College Station, Texas

**Vatche P. Tchakerian,
    Ph.D.**
Professor
Department of
    Geography & Geology
Texas A&M University
College Station, Texas

Teacher Reviewers

Diedre S. Adams
Physical Science Instructor
Science Department
West Vigo Middle School
West Terre Haute,
    Indiana

Laura Buchanan
*Science Teacher and
    Department Chairperson*
Corkran Middle School
Glen Burnie, Maryland

Randy Dye, M.S.
*Middle School Science
    Department Head*
Earth Science
Wood Middle School
Waynesville School
    District #6, Missouri

Laura Kitselman
*Science Teacher and
    Coordinator*
Loudoun Country Day
    School
Leesburg, Virginia

Sally M. Lesley
ESL Science Teacher
Burnet Middle School
Austin, Texas

Susan H. Robinson
Science Teacher
Oglethorpe County Middle
    School
Lexington, Georgia

Marci L. Stadiem
Department Head
Science Department
Cascade Middle School,
    Highline School District
Seattle, Washington

Lab Development

Kenneth E. Creese
Science Teacher
White Mountain Junior
    High School
Rock Spring, Wyoming

Linda A. Culp
*Science Teacher and
    Department Chair*
Thorndale High School
Thorndale, Texas

Bruce M. Jones
*Science Teacher and
    Department Chair*
The Blake School
Minneapolis, Minnesota

Shannon Miller
Science and Math Teacher
Llano Junior High School
Llano, Texas

Robert Stephen Ricks
Special Services Teacher
Department of Classroom
    Improvement
Alabama State Department
    of Education
Montgomery, Alabama

James J. Secosky
Science Teacher
Bloomfield Central School
Bloomfield, New York

Lab Testing

Barry L. Bishop
*Science Teacher and
    Department Chair*
San Rafael Junior High
Ferron, Utah

Janel Guse
*Science Teacher and
    Department Chair*
West Central Middle School
Hartford, South Dakota

David Jones
Science Teacher
Andrew Jackson Middle
    School
Cross Lanes, West Virginia

Michael E. Kral
Science Teacher
West Hardin Middle School
Cecilia, Kentucky

Bert Sherwood
Science Teacher
Socorro Middle School
El Paso, Texas

David M. Sparks
Science Teacher
Redwater Junior High
    School
Redwater, Texas

Larry Tackett
*Science Teacher and
    Department Chair*
Andrew Jackson Middle
    School
Cross Lanes, West Virginia

Answer Checking

Catherine Podeszwa
Duluth, Minnesota

Feature Development

Katy Z. Allen
Hatim Belyamani
John A. Benner
David Bradford
Jennifer Childers
Mickey Coakley
Susan Feldkamp
Jane Gardner
Erik Hahn
Christopher Hess
Deena Kalai
Charlotte W. Luongo, MSc
Michael May
Persis Mehta, Ph.D.
Eileen Nehme, MPH
Catherine Podeszwa
Dennis Rathnaw
Daniel B. Sharp
John Stokes
April Smith West
Molly F. Wetterschneider

G Earth's Changing Surface

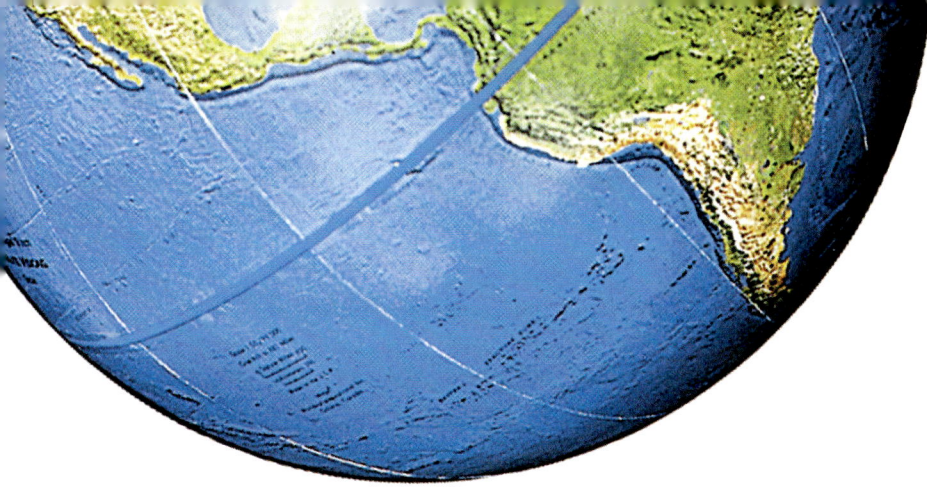

Labs and Activities

vii

How to Use Your Textbook

Your Roadmap for Success with Holt Science and Technology

Reading Warm-Up

A Reading Warm-Up at the beginning of every section provides you with the section's objectives and key terms. The objectives tell you what you'll need to know after you finish reading the section.

Key terms are listed for each section. Learn the definitions of these terms because you will most likely be tested on them. Each key term is highlighted in the text and is defined at point of use and in the margin. You can also use the glossary to locate definitions quickly.

STUDY TIP Reread the objectives and the definitions to the key terms when studying for a test to be sure you know the material.

Get Organized

A Reading Strategy at the beginning of every section provides tips to help you organize and remember the information covered in the section. Keep a science notebook so that you are ready to take notes when your teacher reviews the material in class. Keep your assignments in this notebook so that you can review them when studying for the chapter test.

SECTION 3

From Bedrock to Soil

Most plants need soil to grow. But what exactly is soil? Where does it come from?

READING WARM-UP

Objectives
- Describe the source of soil.
- Explain how the different properties of soil affect plant growth.
- Describe how various climates affect soil.

Terms to Learn

soil
parent rock
bedrock
soil texture
soil structure
humus
leaching

READING STRATEGY

Prediction Guide Before you read this section, write the title of each heading in this section. Next, under each heading, write what you think you will learn.

The Source of Soil

To a scientist, **soil** is a loose mixture of small mineral fragments, organic material, water, and air that can support the growth of vegetation. But not all soils are the same. Because soils are made from weathered rock fragments, the type of soil that forms depends on the type of rock that weathers. The rock formation that is the source of mineral fragments in the soil is called **parent rock.**

Bedrock is the layer of rock beneath soil. In this case, the bedrock is the parent rock because the soil above it formed from the bedrock below. Soil that remains above its parent rock is called *residual soil.*

Soil can be blown or washed away from its parent rock. This soil is called *transported soil.* **Figure 1** shows one way that soil is moved from one place to another. Both wind and the movement of glaciers are also responsible for transporting soil.

✓ *Reading Check* What is soil formed from? (*See the Appendix for answers to Reading Checks.*)

soil a loose mixture of rock fragments, organic material, water, and air that can support the growth of vegetation

parent rock a rock formation that is the source of soil

bedrock the layer of rock beneath soil

Figure 1 *Transported soil may be moved long distances from its parent rock by rivers, such as this one.*

288   Chapter 10   Weathering and Soil Formation

Be Resourceful—Use the Web

Internet Connect

boxes in your textbook take you to resources that you can use for science projects, reports, and research papers. Go to scilinks.org, and type in the SciLinks code to get information on a topic.

Visit go.hrw.com

Find worksheets, **Current Science**® magazine articles online, and other materials that go with your textbook at **go.hrw.com.** Click on the textbook icon and the table of contents to see all of the resources for each chapter.

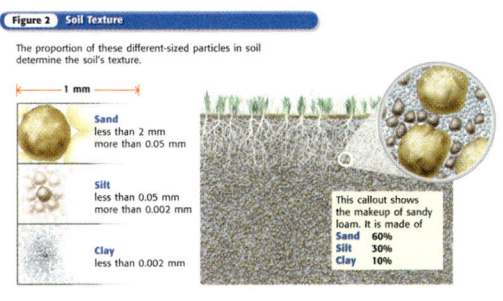

Figure 2 Soil Texture

The proportion of these different-sized particles in soil determine the soil's texture.

| 1 mm |

Sand
less than 2 mm
more than 0.05 mm

Silt
less than 0.05 mm
more than 0.002 mm

Clay
less than 0.002 mm

This callout shows the makeup of sandy loam. It is made of
Sand 60%
Silt 30%
Clay 10%

Soil Properties

Some soils are great for growing plants. Other soils can't support the growth of plants. To better understand soil, you will next learn about its properties, such as soil texture, soil structure, and soil fertility.

Soil Texture and Soil Structure

Soil is made of different-sized particles. These particles can be as large as 2 mm, such as sand. Other particles can be too small to see without a microscope. **Soil texture** is the soil quality that is based on the proportions of soil particles. **Figure 2** shows the soil texture for one type of soil.

Soil texture affects the soil's consistency. Consistency describes a s⸱⸱⸱⸱ility ⸱⸱⸱⸱⸱⸱⸱⸱ ⸱⸱⸱⸱ ⸱⸱⸱⸱⸱⸱ ⸱⸱⸱⸱ ⸱⸱⸱⸱ ⸱⸱⸱⸱ ing. For exa⸱⸱⸱ clay can be ⸱⸱

Soil text⸱⸱ move throu⸱⸱ roots witho⸱⸱

Water an⸱⸱ soil structure⸱ Soil particles of soil parti⸱⸱ of soil can ⸱ affects soil r⸱⸱

soil texture the soil quality that is based on the proportions of soil particles

soil structure the arrangement of soil particles

Arctic Climates

Arctic areas have so little precipitation that they are like cold deserts. In arctic climates, as in desert climates, chemical weathering occurs very slowly. So, soil formation also occurs slowly. Slow soil formation is why soil in arctic areas, as shown in **Figure 7,** is thin and unable to support many plants.

Arctic climates also have low soil temperatures. At low temperatures, decomposition of plants and animals happens more slowly or stops completely. Slow decomposition limits the amount of humus in the soil, which limits the nutrients available. These nutrients are necessary for plant growth.

Figure 7 Arctic soils, such as the soil along Denali Highway, in Alaska, cannot support lush vegetation.

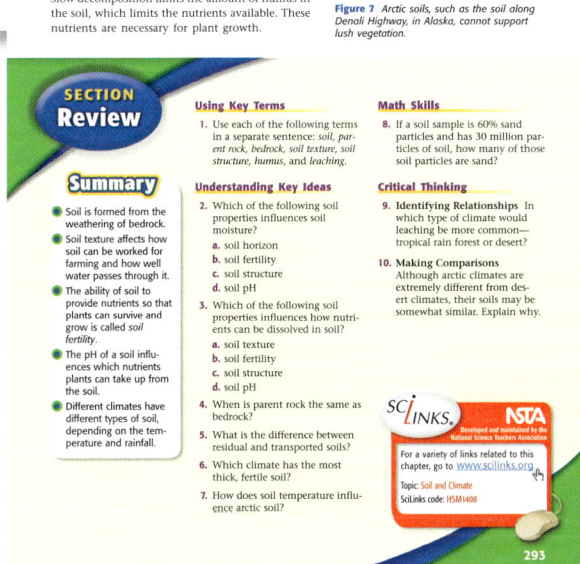

SECTION Review

Summary

- Soil is formed from the weathering of bedrock.
- Soil texture affects how soil can be worked for farming and how well water passes through it.
- The ability of soil to provide nutrients so that plants can survive and grow is called *soil fertility.*
- The pH of a soil influences which nutrients plants can take up from the soil.
- Different climates have different types of soil, depending on the temperature and rainfall.

Using Key Terms

1. Use each of the following terms in a separate sentence: *soil, parent rock, bedrock, soil texture, soil structure, humus,* and *leaching.*

Understanding Key Ideas

2. Which of the following soil properties influences soil moisture?
   a. soil horizon
   b. soil fertility
   c. soil structure
   d. soil pH

3. Which of the following soil properties influences how nutrients can be dissolved in soil?
   a. soil texture
   b. soil fertility
   c. soil structure
   d. soil pH

4. When is parent rock the same as bedrock?

5. What is the difference between residual and transported soils?

6. Which climate has the most thick, fertile soil?

7. How does soil temperature influence arctic soil?

Math Skills

8. If a soil sample is 60% sand particles and has 30 million particles of soil, how many of those soil particles are sand?

Critical Thinking

9. **Identifying Relationships** In which type of climate would leaching be more common—tropical rain forest or desert?

10. **Making Comparisons** Although arctic climates are extremely different from desert climates, their soils may be somewhat similar. Explain why.

SCLINKS. **NSTA**
Developed and maintained by the National Science Teachers Association

For a variety of links related to this chapter, go to www.scilinks.org

Topic: Soil and Climate
SciLinks code: HSM1408

293

Use the Illustrations and Photos

Art shows complex ideas and processes. Learn to analyze the art so that you better understand the material you read in the text.

Tables and graphs display important information in an organized way to help you see relationships.

A picture is worth a thousand words. Look at the photographs to see relevant examples of science concepts that you are reading about.

Answer the Section Reviews

Section Reviews test your knowledge of the main points of the section. Critical Thinking items challenge you to think about the material in greater depth and to find connections that you infer from the text.

STUDY TIP When you can't answer a question, reread the section. The answer is usually there.

Do Your Homework

Your teacher may assign worksheets to help you understand and remember the material in the chapter.

STUDY TIP Don't try to answer the questions without reading the text and reviewing your class notes. A little preparation up front will make your homework assignments a lot easier. Answering the items in the Chapter Review will help prepare you for the chapter test.

Holt Online Learning

Visit Holt Online Learning

If your teacher gives you a special password to log onto the Holt Online Learning site, you'll find your complete textbook on the Web. In addition, you'll find some great learning tools and practice quizzes. You'll be able to see how well you know the material from your textbook.

CNN Student News™

Visit CNN Student News

You'll find up-to-date events in science at **cnnstudentnews.com.**

SAFETY FIRST!

Exploring, inventing, and investigating are essential to the study of science. However, these activities can also be dangerous. To make sure that your experiments and explorations are safe, you must be aware of a variety of safety guidelines. You have probably heard of the saying, "It is better to be safe than sorry." This is particularly true in a science classroom where experiments and explorations are being performed. Being uninformed and careless can result in serious injuries. Don't take chances with your own safety or with anyone else's.

The following pages describe important guidelines for staying safe in the science classroom. Your teacher may also have safety guidelines and tips that are specific to your classroom and laboratory. Take the time to be safe.

Safety Rules!

Start Out Right

Always get your teacher's permission before attempting any laboratory exploration. Read the procedures carefully, and pay particular attention to safety information and caution statements. If you are unsure about what a safety symbol means, look it up or ask your teacher. You cannot be too careful when it comes to safety. If an accident does occur, inform your teacher immediately regardless of how minor you think the accident is.

If you are instructed to note the odor of a substance, wave the fumes toward your nose with your hand. Never put your nose close to the source.

Safety Symbols

All of the experiments and investigations in this book and their related worksheets include important safety symbols to alert you to particular safety concerns. Become familiar with these symbols so that when you see them, you will know what they mean and what to do. It is important that you read this entire safety section to learn about specific dangers in the laboratory.

Eye protection

Clothing protection

Hand safety

Heating safety

Electric safety

Chemical safety

Animal safety

Sharp object

Plant safety

Eye Safety

Wear safety goggles when working around chemicals, acids, bases, or any type of flame or heating device. Wear safety goggles any time there is even the slightest chance that harm could come to your eyes. If any substance gets into your eyes, notify your teacher immediately and flush your eyes with running water for at least 15 minutes. Treat any unknown chemical as if it were a dangerous chemical. Never look directly into the sun. Doing so could cause permanent blindness.

Avoid wearing contact lenses in a laboratory situation. Even if you are wearing safety goggles, chemicals can get between the contact lenses and your eyes. If your doctor requires that you wear contact lenses instead of glasses, wear eye-cup safety goggles in the lab.

Safety Equipment

Know the locations of the nearest fire alarms and any other safety equipment, such as fire blankets and eyewash fountains, as identified by your teacher, and know the procedures for using the equipment.

Neatness

Keep your work area free of all unnecessary books and papers. Tie back long hair, and secure loose sleeves or other loose articles of clothing, such as ties and bows. Remove dangling jewelry. Don't wear open-toed shoes or sandals in the laboratory. Never eat, drink, or apply cosmetics in a laboratory setting. Food, drink, and cosmetics can easily become contaminated with dangerous materials.

Certain hair products (such as aerosol hair spray) are flammable and should not be worn while working near an open flame. Avoid wearing hair spray or hair gel on lab days.

Sharp/Pointed Objects

Use knives and other sharp instruments with extreme care. Never cut objects while holding them in your hands. Place objects on a suitable work surface for cutting.

Be extra careful when using any glassware. When adding a heavy object to a graduated cylinder, tilt the cylinder so that the object slides slowly to the bottom.

Heat

Wear safety goggles when using a heating device or a flame. Whenever possible, use an electric hot plate as a heat source instead of using an open flame. When heating materials in a test tube, always angle the test tube away from yourself and others. To avoid burns, wear heat-resistant gloves whenever instructed to do so.

Electricity

Be careful with electrical cords. When using a microscope with a lamp, do not place the cord where it could trip someone. Do not let cords hang over a table edge in a way that could cause equipment to fall if the cord is accidentally pulled. Do not use equipment with damaged cords. Be sure that your hands are dry and that the electrical equipment is in the "off" position before plugging it in. Turn off and unplug electrical equipment when you are finished.

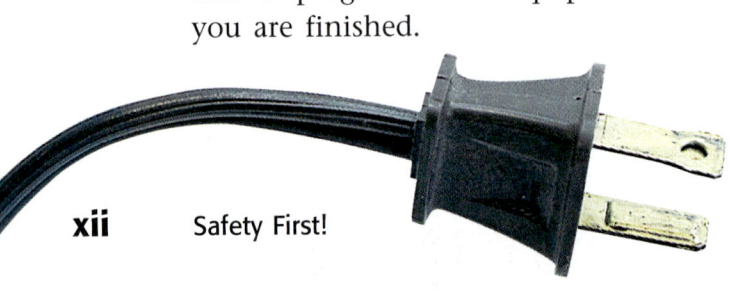

Chemicals

Wear safety goggles when handling any potentially dangerous chemicals, acids, or bases. If a chemical is unknown, handle it as you would a dangerous chemical. Wear an apron and protective gloves when you work with acids or bases or whenever you are told to do so. If a spill gets on your skin or clothing, rinse it off immediately with water for at least 5 minutes while calling to your teacher.

Never mix chemicals unless your teacher tells you to do so. Never taste, touch, or smell chemicals unless you are specifically directed to do so. Before working with a flammable liquid or gas, check for the presence of any source of flame, spark, or heat.

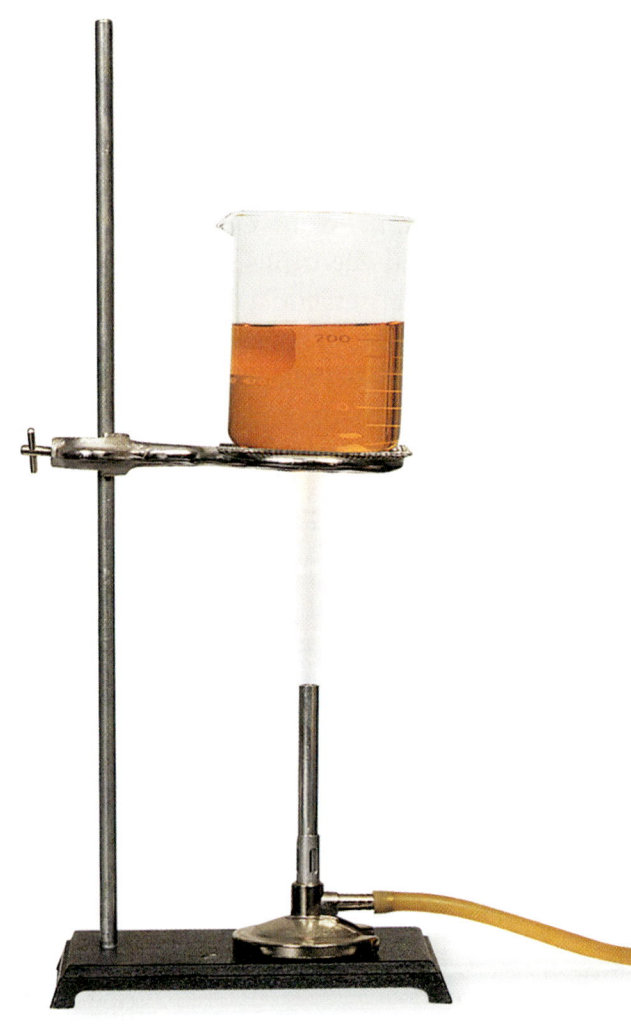

Animal Safety

Always obtain your teacher's permission before bringing any animal into the school building. Handle animals only as your teacher directs. Always treat animals carefully and respectfully. Wash your hands thoroughly after handling any animal.

Plant Safety

Do not eat any part of a plant or plant seed used in the laboratory. Wash your hands thoroughly after handling any part of a plant. When in nature, do not pick any wild plants unless your teacher instructs you to do so.

Glassware

Examine all glassware before use. Be sure that glassware is clean and free of chips and cracks. Report damaged glassware to your teacher. Glass containers used for heating should be made of heat-resistant glass.

Maps as Models of the Earth
Chapter Planning Guide

Compression guide:
To shorten instruction because of time limitations, omit the Chapter Lab.

OBJECTIVES	LABS, DEMONSTRATIONS, AND ACTIVITIES	TECHNOLOGY RESOURCES
PACING • 90 min pp. 2–9 **Chapter Opener**	SE **Start-up Activity**, p. 3 `GENERAL`	OSP **Parent Letter** ■ `GENERAL` CD **Student Edition on CD-ROM** CD **Guided Reading Audio CD** ■ TR **Chapter Starter Transparency*** VID Brain Food Video Quiz
Section 1 You Are Here • Explain how a magnetic compass can be used to find directions on Earth. • Explain the difference between true north and magnetic north. • Compare latitude and longitude. • Explain how latitude and longitude are used to locate places on Earth.	TE **Group Activity** Using Directions, p. 4 `GENERAL` SE **School-to-Home Activity** Columbus's Voyage, p. 5 `GENERAL` SE **Quick Lab** Making a Compass, p. 6 ◆ `GENERAL` CRF **Datasheet for Quick Lab*** TE **Group Activity** Finding True North, p. 7 `GENERAL` SE **Connection to Social Studies** Global Addresses, p. 8 `GENERAL` SE **Skills Practice Lab** Round or Flat?, p. 22 ◆ `GENERAL` CRF **Datasheet for Chapter Lab***	CRF **Lesson Plans*** TR **Bellringer Transparency*** TR **The North and South Poles*** TR **Lines of Longitude; Lines of Latitude*** VID **Lab Videos for Earth Science** CD **Science Tutor**
PACING • 45 min pp. 10–17 **Section 2 Mapping the Earth's Surface** • Explain why maps of the Earth show distortion. • Describe four types of map projections. • Identify five pieces of information that should be shown on a map. • Describe four methods modern mapmakers use to make accurate maps.	SE **Science in Action** Math, Social Studies, and Language Arts Activities, pp. 28–29 `GENERAL` TE **Group Activity** Comparing Map Projections, p. 13 `BASIC` TE **Connection Activity** Math, p. 14 `GENERAL` TE **Connection Activity** Language Arts, p. 15 `ADVANCED` TE **Activity** Remote-Sensing Technology, p. 15 `GENERAL`	CRF **Lesson Plans*** TR **Bellringer Transparency*** TR **Cylindrical Projection; Conic Projection; Azimuthal Projection*** TR *LINK TO PHYSICAL SCIENCE* The Electromagnetic Spectrum*** TE **Internet Activity**, p. 11 CRF **SciLinks Activity*** `GENERAL` CD **Science Tutor**
PACING • 45 min pp. 18–21 **Section 3 Topographic Maps** • Explain how contour lines show elevation and landforms on a map. • Explain how the relief of an area determines the contour interval used on a map. • List the rules of contour lines.	TE **Activity** Investigate Your Area, p. 18 ◆ `GENERAL` TE **Group Activity** Model Terrain, p. 19 `BASIC` SE **Skills Practice Lab** Topographic Tuber, p. 92 `GENERAL` CRF **Datasheet for LabBook*** LB **Inquiry Labs** Looking for Buried Treasure* `ADVANCED` LB **Long-Term Projects & Research Ideas** Globe Trotting* `ADVANCED`	CRF **Lesson Plans*** CRF **Bellringer Transparency*** CD **Science Tutor**

PACING • 90 min

CHAPTER REVIEW, ASSESSMENT, AND STANDARDIZED TEST PREPARATION

CRF **Vocabulary Activity*** `GENERAL`
SE **Chapter Review**, pp. 24–25 `GENERAL`
CRF **Chapter Review*** ■ `GENERAL`
CRF **Chapter Tests A*** ■ `GENERAL`, **B*** `ADVANCED`, **C*** `SPECIAL NEEDS`
SE **Standardized Test Preparation**, pp. 26–27 `GENERAL`
CRF **Standardized Test Preparation*** `GENERAL`
CRF **Performance-Based Assessment*** `GENERAL`
OSP **Test Generator** `GENERAL`
CRF **Test Item Listing*** `GENERAL`

Online and Technology Resources

Visit **go.hrw.com** for a variety of free resources related to this textbook. Enter the keyword **HZ5MAP**.

Holt Online Learning

Students can access interactive problem-solving help and active visual concept development with the *Holt Science and Technology* Online Edition available at **www.hrw.com**.

 Guided Reading Audio CD
Also in Spanish

A direct reading of each chapter for auditory learners, reluctant readers, and Spanish-speaking students.

 Science Tutor CD-ROM

Excellent for remediation and test practice.

SKILLS DEVELOPMENT RESOURCES | SECTION REVIEW AND ASSESSMENT | CORRELATIONS

SE Pre-Reading Activity, p. 2 `GENERAL`
OSP Science Puzzlers, Twisters & Teasers `GENERAL`

National Science Education Standards

UCP 2, 3, 5; ST 1, 2; SAI 1; SPSP 5; HNS 1, 3

CRF Directed Reading A* ■ `BASIC`, B* `SPECIAL NEEDS`
CRF Vocabulary and Section Summary* ■ `GENERAL`
SE Reading Strategy Reading Organizer, p. 4 `GENERAL`
TE Reading Strategy Prediction Guide, p. 6 `GENERAL`
TE Inclusion Strategies, p. 7 ◆
CRF Reinforcement Worksheet Where on Earth?* `BASIC`

SE Reading Checks, pp. 5, 6, 8 `GENERAL`
TE Reteaching, p. 8 `BASIC`
TE Quiz, p. 8 `GENERAL`
TE Alternative Assessment, p. 8 `GENERAL`
SE Section Review,* p. 9 ■ `GENERAL`
CRF Section Quiz* ■ `GENERAL`

UCP 2, 3, 5; SAI 1; ST 1, 2; SPSP 5; HNS 1, 3

Chapter Lab: UCP 2, 3; SAI 1; HNS 3

CRF Directed Reading A* ■ `BASIC`, B* `SPECIAL NEEDS`
CRF Vocabulary and Section Summary* ■ `GENERAL`
SE Reading Strategy Discussion, p. 10 `GENERAL`
SE Connection to Social Studies Mapmaking and Ship Navigation, p. 12 `GENERAL`
TE Reading Strategy Making an Outline, p. 11 `BASIC`
TE Reading Strategy Mnemonics, p. 12 `GENERAL`
TE Inclusion Strategies, p. 14 ◆
MS Math Skills for Science Using Proportions and Cross-Multiplication* `GENERAL`
CRF Critical Thinking Shaping the World* `ADVANCED`

SE Reading Checks, pp. 10, 13, 14, 16 `GENERAL`
TE Homework, p. 12 `GENERAL`
TE Homework, p. 13 `ADVANCED`
TE Reteaching, p. 16 `BASIC`
TE Quiz, p. 16 `GENERAL`
TE Alternative Assessment, p. 16 `GENERAL`
SE Section Review,* p. 17 ■ `GENERAL`
CRF Section Quiz* ■ `GENERAL`

UCP 2, 3, 5; SAI 1; ST 2; SPSP 5; HNS 3

CRF Directed Reading A* ■ `BASIC`, B* `SPECIAL NEEDS`
CRF Vocabulary and Section Summary* ■ `GENERAL`
SE Reading Strategy Paired Summarizing, p. 18 `GENERAL`
SE Connection to Oceanography Mapping the Ocean Floor, p. 19 `GENERAL`
SE Connection to Environmental Science Endangered Species, p. 21 `GENERAL`
CRF Reinforcement Worksheet Interpreting a Topographic Map* `BASIC`
MS Math Skills for Science Mapping and Surveying `GENERAL`

SE Reading Checks, p. 19 `GENERAL`
TE Homework, p. 19 `ADVANCED`
TE Reteaching, p. 20 `BASIC`
TE Quiz, p. 20 `GENERAL`
TE Alternative Assessment, p. 20 `GENERAL`
SE Section Review,* p. 21 ■ `GENERAL`
CRF Section Quiz* ■ `GENERAL`

UCP 2, 3, 5; SAI 1; ST 2; SPSP 2, 5; HNS 1

LabBook: UCP 2, 3; SAI 1; ST 1

One-Stop Planner® CD-ROM

This CD-ROM package includes:
• Lab Materials QuickList Software
• Holt Calendar Planner
• Customizable Lesson Plans
• Printable Worksheets
• ExamView® Test Generator
• Interactive Teacher Edition
• Holt PuzzlePro® Resources
• Holt PowerPoint® Resources

www.scilinks.org

Maintained by the **National Science Teachers Association.** See Chapter Enrichment pages for a complete list of topics.

Current Science®

Check out *Current Science* articles and activities by visiting the HRW Web site at **go.hrw.com.** Just type in the keyword **HZ5CS02T.**

 Classroom Videos

• **Lab Videos** demonstrate the chapter lab.
• **Brain Food Video Quizzes** help students review the chapter material.
• **CNN Videos** bring science into your students' daily life.

Chapter Resources

Visual Resources

CHAPTER STARTER TRANSPARENCY

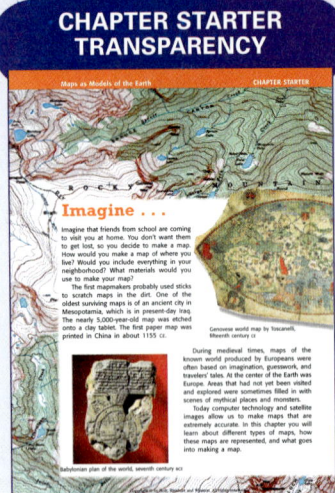

BELLRINGER TRANSPARENCIES

Section: You Are Here
Everyone uses maps. Some maps are very formal with lots of detail, compass points, and drawn to perfect scale. Some maps are just useful memories you use to know the quickest way to class or to the cafeteria. Draw a map from your house to one of your favorite places. Clearly label all landmarks, and include information that might be useful to someone using the map.

Draw your map in your **science journal**.

Section: Mapping the Earth's Surface
Compare the world map, the state map, and the community map. Make a chart of the similarities and differences between each map. You might use the world map to plan a big trip, or to look up a place you heard of on the news. Can you think of three uses for each kind of map? Can you think of three improvements you could make to each of these maps?

Record your answers in your **science journal**.

TEACHING TRANSPARENCIES

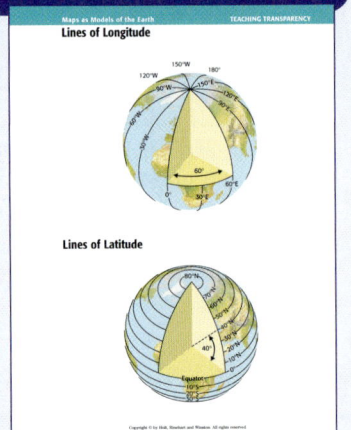

TEACHING TRANSPARENCIES

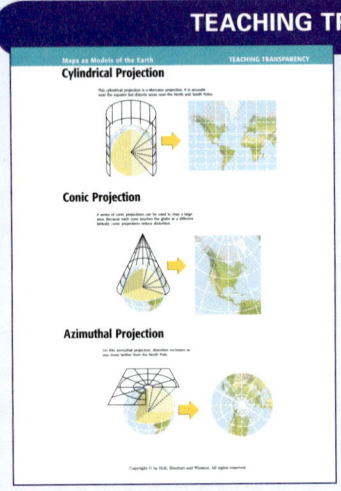

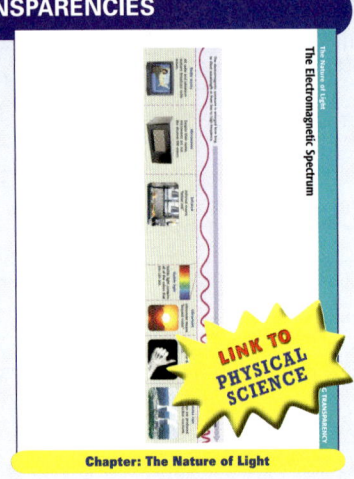

CONCEPT MAPPING TRANSPARENCY

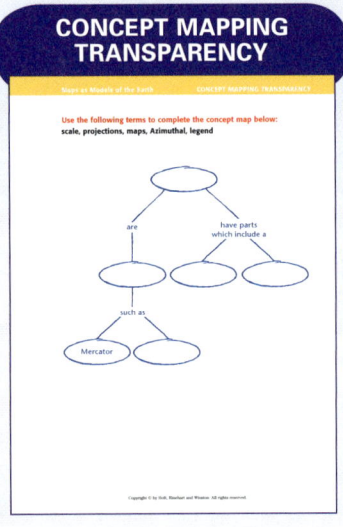

Planning Resources

LESSON PLANS

PARENT LETTER

TEST ITEM LISTING

One-Stop Planner® CD-ROM

This CD-ROM includes all of the resources shown here and the following time-saving tools:

- *Lab Materials QuickList Software*
- *Customizable lesson plans*
- *Holt Calendar Planner*
- *The powerful ExamView® Test Generator*

Meeting Individual Needs

DIRECTED READING A
BASIC — ALSO IN SPANISH

DIRECTED READING B
SPECIAL NEEDS

VOCABULARY ACTIVITY
GENERAL

VOCABULARY AND SECTION SUMMARY
GENERAL — ALSO IN SPANISH

REINFORCEMENT
BASIC

CRITICAL THINKING
ADVANCED

SCILINKS ACTIVITY
GENERAL

SCIENCE PUZZLERS, TWISTERS & TEASERS
GENERAL

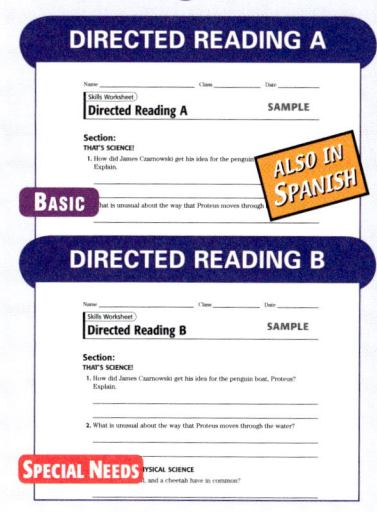
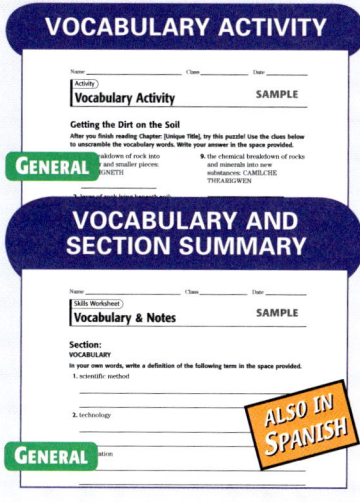
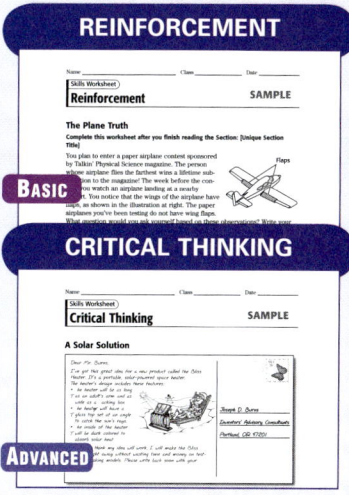

Labs and Activities

LONG-TERM PROJECTS & RESEARCH IDEAS
ADVANCED

INQUIRY LABS
ADVANCED

DATASHEETS FOR QUICK LABS

DATASHEETS FOR CHAPTER LABS

DATASHEETS FOR LABBOOK

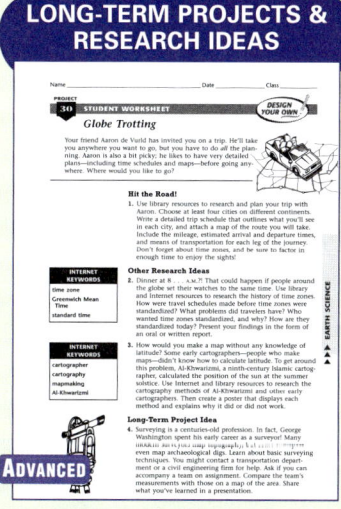
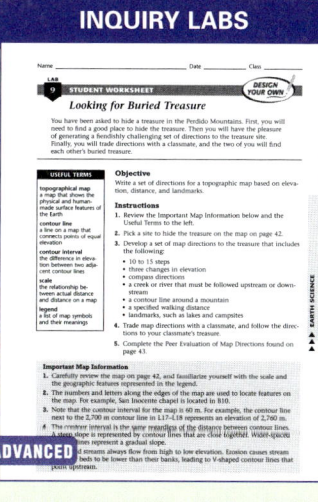
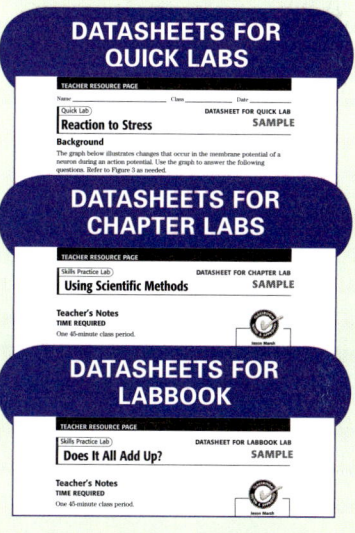

Review and Assessments

SECTION QUIZ
GENERAL — ALSO IN SPANISH

SECTION REVIEW
GENERAL — ALSO IN SPANISH

CHAPTER REVIEW
GENERAL — ALSO IN SPANISH

CHAPTER TEST A
GENERAL — ALSO IN SPANISH

CHAPTER TEST B
ADVANCED

CHAPTER TEST C
SPECIAL NEEDS

STANDARDIZED TEST PREPARATION
GENERAL

PERFORMANCE-BASED ASSESSMENT
GENERAL

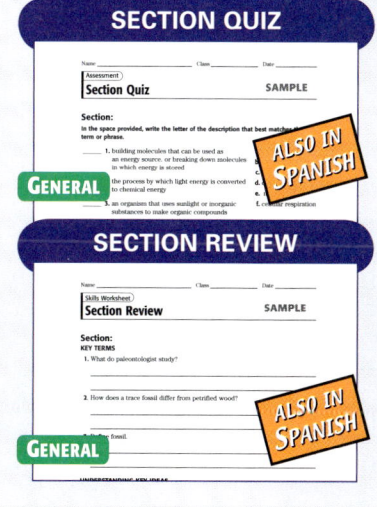
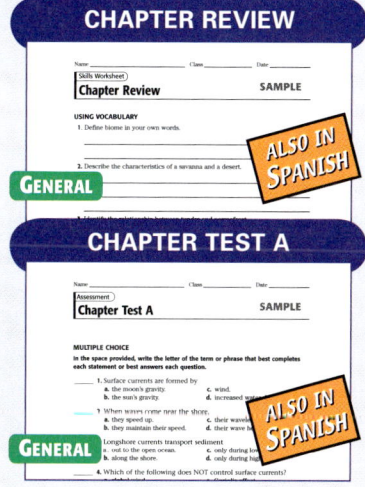
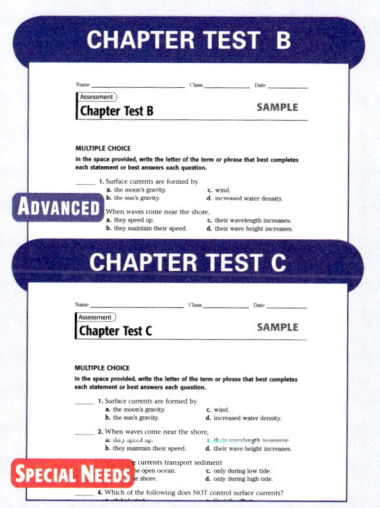
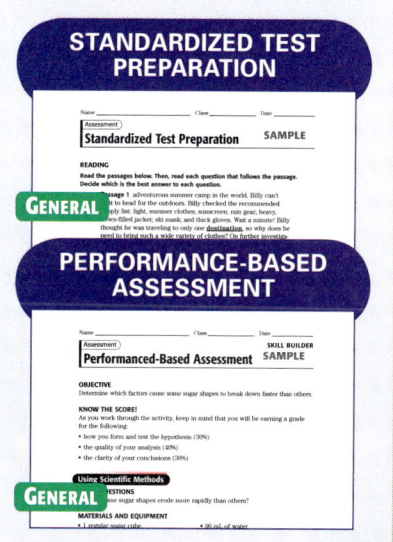

Chapter Enrichment

This Chapter Enrichment provides relevant and interesting information to expand and enhance your presentation of the chapter material.

Section 1

You Are Here
Longitude

● Because Earth rotates 360° every 24 h, it turns 15° every hour. Therefore, longitude can be determined at any place on the globe if the local time and the time at the prime meridian are known. Before the mid-18th century, the unreliability of clocks—especially those aboard ships, where motion, temperature variation, and moisture could wreak havoc with a timepiece's workings—thwarted calculations of longitude. Many shipwrecks occurred because the ship captains could not accurately calculate their location.

● In 1707, inaccurate longitudinal information caused four ships in a British fleet to run aground, and 2,000 sailors died. The British Parliament addressed the problem by offering a large reward to anyone who could develop a method to accurately calculate longitude within half a degree. John Harrison (1693–1776), a self-taught clockmaker, developed a chronometer that remained accurate on rough seas, and he won the prize in 1763. More than 200 years later, astronaut Neil Armstrong gave credit to Harrison for the role he played in enabling exploration of Earth and in inspiring future generations to venture toward exploration of the moon.

Section 2

Mapping the Earth's Surface
Gerardus Mercator

● Gerardus Mercator was born Gerhard Kremer in 1512 in Rupelmonde, Flanders (present-day Belgium). At age 24, Mercator was a highly skilled engraver, calligrapher, and scientific-instrument maker. With two of his teachers, he made the first globe of the Earth

in 1536–1537. A true Renaissance man, Mercator was a highly esteemed cartographer, who also published a treatise on italic lettering, designed a grammar-school curriculum, taught mathematics, and conducted genealogical research for his patron, Duke Wilhelm of Cleve. He even attempted to write a chronology of the history of the world from the formation of Earth to 1568.

Is That a Fact!

◆ In 1544, Gerardus Mercator was imprisoned on charges of treason. Apparently, his frequent absences from Flanders to gather map data aroused the suspicions of authorities. He remained imprisoned for 7 months before his friends succeeded in clearing his name.

The Global Positioning System

● During the 1970s, the U.S. Department of Defense developed the Global Positioning System (GPS) for use in aircraft navigation and missile guidance. The system uses a network of satellites that continuously transmit positioning information to receivers on Earth. The distance between a receiver and at least four satellites is used to compute the latitude and longitude coordinates of the receiver's position.

● In 1983, the system was made available to the public and has since been used for land, sea, and air navigation, surveying, and geophysical exploration. The system can be highly accurate, making it possible to determine a position within less than 1 m. As the prices of some of the receivers have plummeted, the use of the GPS for recreational activities, such as boating, hiking, and hunting, has increased. Most receivers used by the public for recreational purposes are accurate only to about 10 m.

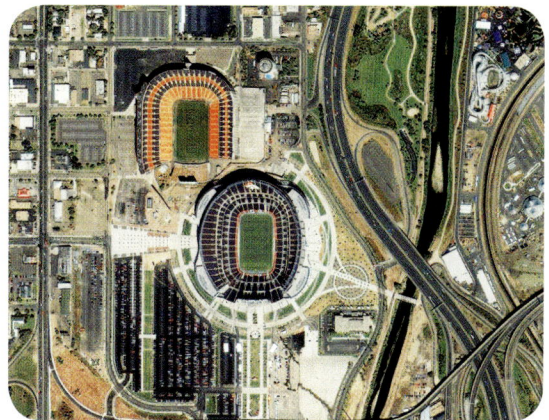

Landsat

- Since the Landsat program began in 1972, a number of satellites have been deployed that carry remote-sensing equipment. The equipment is designed to detect radiation in different bands of the electromagnetic spectrum. The most recent satellites, *Landsat 4, 5,* and *7,* orbit Earth from pole to pole every 16 days. Landsat data are particularly useful for thematic mapping. For example, data from the blue-green spectral region are useful for distinguishing between coniferous and deciduous plants, and data from the thermal infrared range supply information about soil moisture.

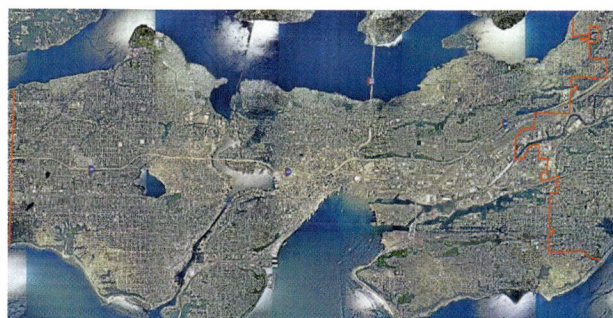

Section 3

Topographic Maps
Inuit Relief Maps

- The Inuit of Baffin Island were skilled mapmakers. They made permanent relief maps by carving coastal features into pieces of wood and walrus ivory. The Inuit also sewed small pieces of fur or driftwood to sealskin to represent islands. They measured distance on their maps not by miles but by "sleeps." The distance to a hunting ground, for example, would be measured by how many rest stops would be taken before reaching it.

John Wesley Powell (1834–1902)

- American geologist and surveyor John Wesley Powell headed an official expedition to the Grand Canyon in 1871. His purpose was to conduct a topographic survey to map "as broad a belt of country as it was possible" on both sides of the Colorado and Green Rivers. The expedition yielded meticulously detailed topographic maps for an area that had previously been described as the "great unknown." Those maps were instrumental to Powell's appointment in 1881 as director of the U.S. Geological Survey (USGS).

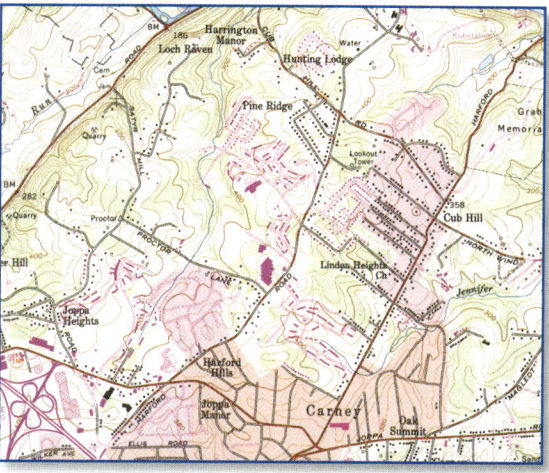

- As director, Powell aimed to create topographic maps for the entire country by using large scales, ranging from 1:250,000 for desert regions to 1:62,500 for densely populated areas. Powell insisted on including data concerning soils, springs, and other natural resources, which he felt were essential for making land-use decisions. The high-quality topographical maps created during Powell's administration set the standard for published topographic maps in the United States for many years to come.

Developed and maintained by the National Science Teachers Association

SciLinks is maintained by the National Science Teachers Association to provide you and your students with interesting, up-to-date links that will enrich your classroom presentation of the chapter.

Visit www.scilinks.org and enter the SciLinks code for more information about the topic listed.

Topic: **Latitude and Longitude**
SciLinks code: **HSM0854**

Topic: **Topographic Maps**
SciLinks code: **HSM1536**

Topic: **Mapmaking**
SciLinks code: **HSM0909**

Overview

This chapter will help students learn basic map-reading skills. Students will also learn how a compass is used to find directions and how lines of latitude and longitude are used to identify points on the Earth's surface.

Assessing Prior Knowledge

Students should be familiar with the following topic:

• scientific models

Identifying Misconceptions

As students learn the material in this chapter, some of them may be confused about how a compass is used to determine direction. Some students may think that a compass can keep a person from getting lost. Stress to students that a compass is usually used in conjunction with a map. Also, some students may be confused about the distortion that occurs when information on a sphere is transferred to a flat surface. Students may not recognize that all maps have some degree of distortion.

Maps as Models of the Earth

About the

No ordinary camera took this picture! In fact, a camera wasn't used at all. This image is a radar image of a mountainous area of Tibet. It was taken from the space shuttle. Radar imaging is a method that scientists use to map areas of the Earth from far above the Earth's surface.

PRE-READING ACTIVITY

 **Three-Panel Flip Chart**
Before you read the chapter, create the FoldNote entitled "Three-Panel Flip Chart" described in the **Study Skills** section of the Appendix. Label the flaps of the three-panel flip chart with "Cylindrical projection," "Conical projection," and "Azimuthal projection." As you read the chapter, write information you learn about each category under the appropriate flap.

Standards Correlations

The following codes indicate the National Science Education Standards that correlate to this chapter. The full text of the standards is at the front of the book.

Chapter Opener
SAI 1; SPSP 5; HNS 1, 3

Section 1 You Are Here
UCP 2, 3, 5; SAI 1; ST 1, 2; SPSP 5; HNS 1, 3

Section 2 Mapping Earth's Surface
UCP 2, 3, 5; SAI 1; ST 2; SPSP 5; HNS 3

Section 3 Topographic Maps
UCP 2, 3, 5; SAI 1; ST 2; SPSP 2, 5, HNS 1; *LabBook:* UCP 2, 3; SAI 1; ST 1

Chapter Lab
UCP 2, 3; SAI 1; ST 1; HNS 1

Chapter Review
UCP 2; SAI 2; HNS 3

Science in Action
ST1; SPSP 5; HNS 1, 3

START-UP **ACTIVITY**

MATERIALS

FOR EACH STUDENT
- computer (optional)
- paper
- pencils, colored

Teacher's Note: Before students start their maps, have them brainstorm a list of school landmarks, and suggest that they use the location of these landmarks as reference points in their maps.

Answers

1. Answers may vary. Accept all reasonable responses.
2. Answers may vary. Accept all reasonable responses.
3. Answers may vary.

START-UP **ACTIVITY**

Follow the Yellow Brick Road

In this activity, you will not only learn how to read a map but you will also make a map that someone else can read.

Procedure

1. Use a **computer drawing program or colored pencils and paper** to draw a map that shows how to get from your classroom to another place in your school, such as the gym. Make sure you include enough information for someone unfamiliar with your school to find his or her way.

2. After you finish drawing your map, switch maps with a partner. Examine your classmate's map, and try to figure out where the map is leading you.

Analysis

1. Is your map an accurate picture of your school? Explain your answer.

2. What could you do to make your map better? What are some limitations of your map?

3. Compare your map with your partner's map. How are your maps alike? How are they different?

Chapter Starter Transparency
Use this transparency to help students begin thinking about the history of mapmaking.

Focus

Overview

This section opens with a discussion of the history of mapmaking. Students learn how to find directions on a globe by using reference points such as the North and South Poles. Students learn how a compass is used to find directions and how true north differs from magnetic north. The section closes with a discussion of lines of latitude and longitude and the way they can be used to locate points on the Earth's surface.

🔔 Bellringer

Ask students to draw a map from their home to one of their favorite places. Have them clearly label all landmarks and include information that might be useful to someone using the map.

Motivate

Group Activity — GENERAL

Using Directions Have students write a description of the route they take as they travel between home and school. Then, have them rewrite the description using cardinal directions. Have pairs of students trade descriptions and use a map to check the accuracy of each other's maps. **LS** Verbal   English Language Learners

READING WARM-UP

Objectives

- Explain how a magnetic compass can be used to find directions on Earth.
- Explain the difference between true north and magnetic north.
- Compare latitude and longitude.
- Explain how latitude and longitude is used to locate places on Earth.

Terms to Learn

map	latitude
true north	equator
magnetic declination	longitude
	prime meridian

READING STRATEGY

Reading Organizer As you read this section, create an outline of the section. Use the headings from the section in your outline.

map a representation of the features of a physical body such as Earth

Figure 1 This map shows what explorers thought the world looked like 1,800 years ago.

You Are Here

Have you ever noticed the curve of the Earth's surface? You probably haven't. When you walk across the Earth, it does not appear to be curved. It looks flat.

Over time, ideas about Earth's shape have changed. Maps reflected how people saw the world and what technology was available. A **map** is a representation of the features of a physical body such as Earth. If you look at Ptolemy's (TAHL uh meez) world map from the second century, as shown in **Figure 1,** you might not know what you are looking at. Today satellites give us more accurate images of the Earth. In this section, you will learn how early scientists knew Earth was round long before pictures from space were taken. You will also learn how to find location and direction on Earth's surface.

What Does Earth Really Look Like?

The Greeks thought of Earth as a sphere almost 2,000 years before Christopher Columbus made his voyage in 1492. The observation that a ship sinks below the horizon as it sails into the distance supported the idea of a spherical Earth. If Earth were flat, the ship would not sink below the horizon.

Eratosthenes (ER uh TAHS thuh NEEZ), a Greek mathematician, wanted to know the size of Earth. In about 240 BCE, he calculated Earth's circumference using math and observations of the sun. There were no satellites or computers back then. We now know his calculation was wrong by only 6,250 km!

CHAPTER RESOURCES

Chapter Resource File

- Lesson Plan
- Directed Reading A **BASIC**
- Directed Reading B **SPECIAL NEEDS**

Technology

Transparencies
- Bellringer
- The North and South Poles

Is That a Fact!

The orientation of maps with north at the top is arbitrary. For many centuries, European maps placed east at the top to emphasize the importance of Jerusalem to the Europeans' faith. The Chinese put south at the top of their maps because nothing to the north held any interest for them.

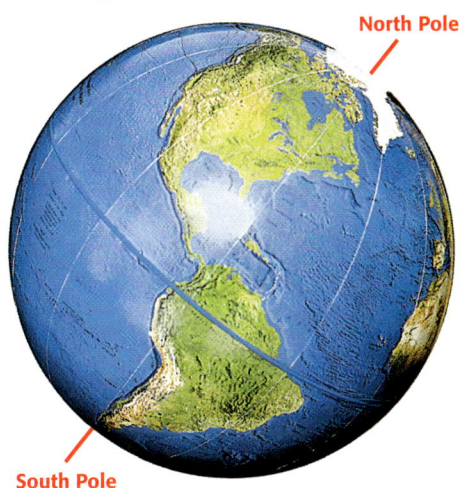

Figure 2 *The North Pole is a good reference point for describing locations in North America.*

North Pole

South Pole

WRITING SKILL **Columbus's Voyage**

Did Christopher Columbus discover that Earth was a sphere only after he completed his voyage in 1492? Or did he know before he left? With a parent, use the Internet or the library to find out more information about Columbus's voyage. Then, write a paragraph describing what you learned.

ACTIVITY

Finding Direction on Earth

When giving directions to your home, you might name a landmark, such as a grocery store, as a reference point. A *reference point* is a fixed place on the Earth's surface from which direction and location can be described.

The Earth is spherical, so it has no top, bottom, or sides for people to use as reference points for determining locations on its surface. However, the Earth does rotate, or spin, on its axis. The Earth's axis is an imaginary line that runs through the Earth. At either end of the axis is a geographic pole. The North and South Poles are used as reference points when describing direction and location on the Earth, as shown in **Figure 2.**

✓ **Reading Check** What is a reference point? (*See the Appendix for answers to Reading Checks.*)

Cardinal Directions

A reference point alone will not help you give good directions. You will need to be able to describe how to get to your home from the reference point. You will need to use the directions north, south, east, and west. These directions are called *cardinal directions.* Using cardinal directions is much more precise than saying "Turn left," "Go straight," or "Turn right." So, you may tell a friend to walk a block north of the gas station to get to your home. To use cardinal directions properly, you will need a compass, shown in **Figure 3.**

Figure 3 *A compass shows the cardinal directions north, south, east and west, as well as combinations of these directions.*

MATERIALS

FOR EACH STUDENT
- compass
- magnet
- needle, sewing, steel
- paper, tissue, 1 cm x 3 cm
- water, in a bowl

Answers

4. Answers may vary. Both compasses should be pointing in the same direction.

5. Answers may vary.

Answer to Reading Check

True north is the direction to the geographic North Pole.

Using a Compass

A magnetic compass will show you which direction is north. A *compass* is a tool that uses the natural magnetism of the Earth to show direction. A compass needle points to the magnetic north pole. Earth has two different sets of poles—the geographic poles and the magnetic poles, as shown in **Figure 4.**

True North and Magnetic Declination

Remember that the Earth's geographic poles are on either end of the Earth's axis. Earth has its own magnetic field, which produces magnetic poles. Earth's magnetic poles are not lined up exactly with Earth's axis. So, there is a difference between the locations of Earth's magnetic and geographic poles. **True north** is the direction to the geographic North Pole. When using a compass, you need to make a correction for the difference between the geographic North Pole and the magnetic north pole. The angle of correction is called **magnetic declination.**

✓ **Reading Check** What is true north?

true north the direction to the geographic North Pole

magnetic declination the difference between the magnetic north and the true north

Making a Compass

1. Do this lab outside. Carefully rub a **steel sewing needle** against a **magnet** in the same direction 40 times.

2. Float a **1 cm × 3 cm piece of tissue paper** in a **bowl of water.**

3. Place the needle in the center of the tissue paper.

4. Compare your compass with a **regular compass.** Are both compasses pointing in the same direction?

5. How would you improve your compass?

Figure 4 *Unlike the geographic poles, which are always in the same place, the magnetic poles have changed location throughout the history of the Earth.*

Magnetic north pole

Geographic North Pole

Geographic South Pole

Magnetic south pole

SCIENCE HUMOR

Before the 17th century, many sailors refused to transport onions and garlic because they believed that these items would destroy a compass's magnetic properties. In 1600, English physician and scientist William Gilbert set out to test this belief. He ate a large quantity of garlic and then belched on a compass needle that he had also rubbed with garlic juice. The compass's magnetic properties remained intact, and Gilbert proved, at least to himself, that the notion was unfounded.

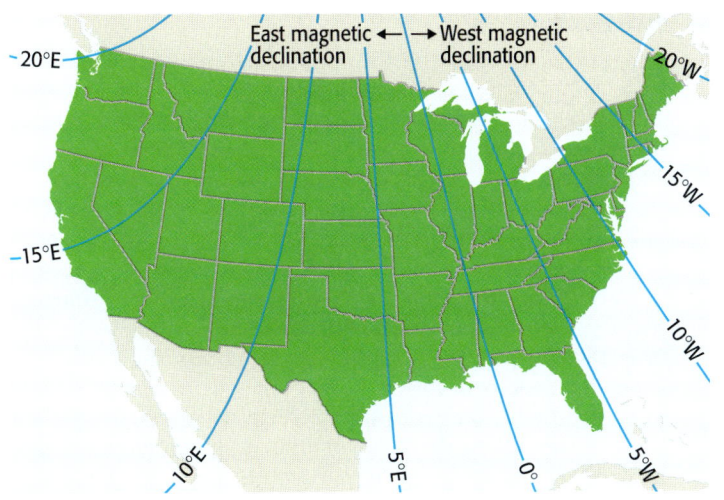

Figure 5 The blue lines on the map connect points that have the same magnetic declination.

Using Magnetic Declination

Magnetic declination is measured in degrees east or west of true north. Magnetic declination has been determined for different points on the Earth's surface. Once you know the declination for your area, you can use a compass to determine true north. This correction is like the correction you would make to the handlebars of a bike with a bent front wheel. You have to turn the handlebars a certain amount to make the bicycle go straight. **Figure 5** shows a map of the magnetic declination of the United States. What is the approximate magnetic declination of your city or town?

Finding Locations on the Earth

All of the houses and buildings in your neighborhood have addresses that give their location. But how would you find the location of something such as a city or an island? These places can be given an "address" using *latitude* and *longitude*. Latitude and longitude are shown by intersecting lines on a globe or map that allow you to find exact locations.

Latitude

Imaginary lines drawn around the Earth parallel to the equator are called lines of latitude, or *parallels*. **Latitude** is the distance north or south from the equator. Latitude is expressed in degrees, as shown in **Figure 6.** The **equator** is a circle halfway between the North and South Poles that divides the Earth into the Northern and Southern Hemispheres. The equator represents 0° latitude. The North Pole is 90° north latitude, and the South Pole is 90° south latitude.

latitude the distance north or south from the equator; expressed in degrees

equator the imaginary circle halfway between the poles that divides the Earth into the Northern and Southern Hemispheres

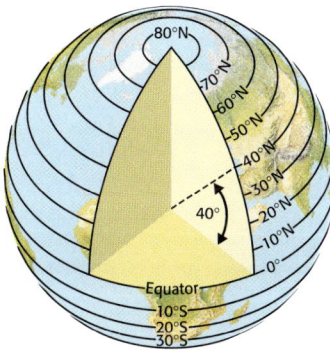

Figure 6 Degrees latitude are a measure of the angle made by the equator and the location on the Earth's surface, as measured from the center of the Earth.

WEIRD SCIENCE

As the Earth rotates, both the geographic North Pole and the magnetic north pole move constantly. The geographic North Pole moves about 6 m on a 435-day cycle. This movement results from a wobble in the Earth's rotation. The magnetic north pole wanders because of changes in Earth's rotating iron core. The magnetic pole is currently moving northwest at an average rate of 10 km per year.

Latitude and Longitude On the board, draw two columns. Label one column "Latitude" and the other column "Longitude." Ask students to call out characteristics of latitude and longitude as you write their answers on the board. Students can copy the information on the board and use it as a study tool.  **Verbal**

Quiz —————————— GENERAL

1. List two reference points that can be used to describe direction and location on Earth. (Sample answer: North Pole, South Pole)

2. What are lines of latitude and lines of longitude? (Lines of latitude are imaginary lines around Earth parallel to the equator that are used to measure a location's distance north or south of the equator. Lines of longitude are imaginary lines that run between the Earth's geographic poles, and they are used to measure a location's distance east or west of the prime meridian.)

Alternative Assessment —————— GENERAL

Planning a Trip Have students use a world map to plan a trip in which they give their various destinations only in degrees of latitude and longitude. Have pairs of students trade their itineraries, and have each student "decode" the other's trip. **L5 Logical**

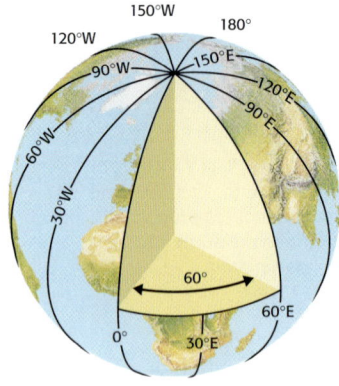

Figure 7 *Degrees longitude are a measure of the angle made by the prime meridian and the location on the Earth's surface, as measured from the center of the Earth.*

longitude the distance east and west from the prime meridian; expressed in degrees

prime meridian the meridian, or line of longitude, that is designated as 0° longitude

Longitude

Lines of longitude, or *meridians*, are imaginary lines that pass through both poles. **Longitude** is the distance east and west from the prime meridian. Like latitude, longitude is expressed in degrees, as shown in **Figure 7**. The **prime meridian** is the line that represents 0° longitude. Unlike lines of latitude, lines of longitude are not parallel. Lines of longitude touch at the poles and are farthest apart at the equator.

Unlike the equator, the prime meridian does not completely circle the globe. The prime meridian runs from the North Pole through Greenwich, England, to the South Pole. The 180° meridian lies on the opposite side of the Earth from the prime meridian. Together, the prime meridian and the 180° meridian divide the Earth into two equal halves—the Eastern and Western Hemispheres. East lines of longitude are found east of the prime meridian, between 0° and 180° longitude. West lines of longitude are found west of the prime meridian, between 0° and 180° longitude.

Using Latitude and Longitude

Points on the Earth's surface can be located by using latitude and longitude. Lines of latitude and lines of longitude cross and form a grid system on globes and maps. This grid system can be used to find locations north or south of the equator and east or west of the prime meridian.

Figure 8 shows you how latitude and longitude can be used to find the location of your state capital. First, locate the star representing your state capital on the appropriate map. Then, use the lines of latitude and longitude closest to your state capital to estimate its approximate latitude and longitude.

 **Which set of imaginary lines are referred to as meridians: lines of latitude or lines of longitude?**

CONNECTION TO Social Studies

Global Addresses You can find the location of any place on Earth by finding the coordinates of the place, or latitude and longitude, on a globe or a map. Using a globe or an atlas, find the coordinates of the following cities:

New York, New York   Madrid, Spain
Sao Paulo, Brazil   Paris, France
Sydney, Australia   Cairo, Egypt

Then, find the latitude and longitude coordinates of your own city. Can you find another city that shares the same latitude as your city? Can you find another city that shares the same longitude?

Answer to Connection to Social Studies
New York, New York: 40°N, 74°W
Sao Paulo, Brazil: 23°S, 47°W
Sydney, Australia: 33°S, 151°E
Madrid, Spain: 40°N, 4°W
Paris, France: 42°N, 2°E
Cairo, Egypt: 30°N, 31°E

Answer to Reading Check
lines of longitude

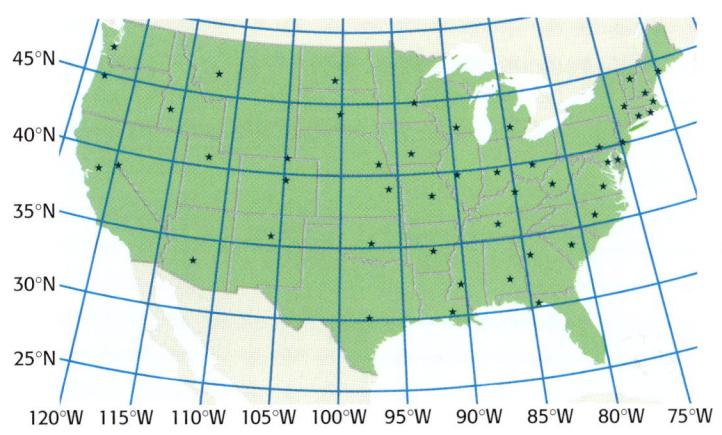

Figure 8 *The grid pattern formed by lines of latitude and longitude allows you to pinpoint any location on the Earth's surface.*

SECTION Review

Summary

- Magnetic compasses are used to find direction on Earth's surface. A compass needle points to the magnetic north pole.

- True north is the direction to the geographic North Pole, which never changes. The magnetic north pole may change over time. Magnetic declination is the difference between true north and magnetic north.

- Latitude and longitude help you find locations on a map or a globe. Lines of latitude run east and west. Lines of longitude run north and south through the poles. These lines cross and form a grid system on globes and maps.

Using Key Terms

1. Use each of the following terms in a separate sentence: *latitude, longitude, equator,* and *prime meridian.*

2. In your own words, write a definition for the term *true north.*

Understanding Key Ideas

3. The geographic poles are
   a. used as reference points when describing direction and location on Earth.
   b. formed because of the Earth's magnetic field.
   c. at either end of the Earth's axis.
   d. Both (a) and (c)

4. How are lines of latitude and lines of longitude alike? How are they different?

5. How can you use a magnetic compass to find directions on Earth?

6. What is the difference between true north and magnetic north?

7. How do lines of latitude and longitude help you find locations on the Earth's surface?

Math Skills

8. The distance between 40°N latitude and 41°N latitude is 69 mi. What is this distance in km? (Hint: 1 km = 0.621 mi)

Critical Thinking

9. **Applying Concepts** While exploring the attic, you find a treasure map. The map shows that the treasure is buried at 97°N and 188°E. Explain why this location is incorrect.

10. **Making Inferences** When using a compass to explore an area, why do you need to know an area's magnetic declination?

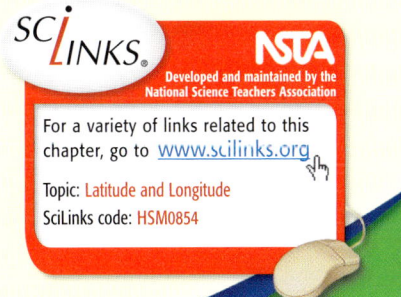

Developed and maintained by the National Science Teachers Association

For a variety of links related to this chapter, go to www.scilinks.org

Topic: Latitude and Longitude
SciLinks code: HSM0854

CHAPTER RESOURCES

Chapter Resource File
- Section Quiz **GENERAL**
- Section Review **GENERAL**
- Vocabulary and Section Summary **GENERAL**
- Reinforcement Worksheet **BASIC**
- Datasheet for Quick Lab

Answers to Section Review

1. Sample answer: Lines of latitude are also referred to as parallels. Lines of longitude are also known as meridians. The equator is an imaginary circle that divides the Earth into the Northern and Southern Hemispheres. The prime meridian is the line that represents 0° longitude.

2. Sample answer: True north is the direction to the Earth's geographic North Pole.

3. d

4. Lines of latitude and lines of longitude are alike because they are both used to describe locations on Earth. Lines of latitude are different from lines of longitude because lines of latitude are parallel to the equator, whereas lines of longitude pass through the poles.

5. Sample answer: A compass needle will point in the direction of the magnetic north pole. You can use this information to find general directions on Earth.

6. True north is the direction to the geographic North Pole. Magnetic north is the direction to the Earth's magnetic north pole. True north and magnetic north have different locations.

7. Lines of latitude and longitude form a grid system that can be used to find locations on the Earth's surface.

8. (69 mi ÷ 0.621 mi) × 1 km = 111.11 km

9. This location is impossible because the greatest measure of latitude is 90° and the greatest measure of longitude is 180°.

10. Because the compass points to magnetic north, it is important to know the magnetic declination at your location. This will help you make corrections to adjust for the difference between true north and magnetic north.

Focus

Overview

In this section, students compare maps and globes as models of the Earth and identify their limitations. Students explore the features of four common map projections. In addition, they discover some of the technological advances that have influenced recent trends in cartography.

🔔 Bellringer

Display a world map, a map of your state, and a map of your community. Have students make a chart in which they list the similarities and differences between each map. Then, have them suggest three uses and three improvements for each map.

Motivate

Discussion —— GENERAL

Map Distortion Have students examine a globe and a Mercator projection of a world map. Tell students that both are representations of Earth. Ask students to find ways in which the two representations differ. Point out the difference in the size and shape of Greenland. Ask students how they might account for the discrepancies. Tell them that in this section they will learn about the difficulties involved in making flat representations of Earth's curved surface. **LS Visual/Verbal**

READING WARM-UP

Objectives

- Explain why maps of the Earth show distortion.
- Describe four types of map projections.
- Identify five pieces of information that should be shown on a map.
- Describe four methods modern map-makers use to make accurate maps.

Terms to Learn

cylindrical projection
conic projection
azimuthal projection
remote sensing

READING STRATEGY

Discussion Read this section silently. Write down questions that you have about this section. Discuss your questions in a small group.

Mapping the Earth's Surface

What do a teddy bear, a toy airplane, and a plastic doll have in common besides being toys? They are all models that represent real things.

Scientists also use models to represent real things, but their models are not toys. Globes and maps are examples of models that scientists use to study the Earth's surface.

Because a globe is a sphere, a globe is the most accurate model of the Earth. A globe accurately shows the sizes and shapes of the continents and oceans in relation to one another. But a globe is not always the best model to use when studying the Earth's surface. A globe is too small to show many details, such as roads and rivers. It is much easier to show details on maps. But how do you show the Earth's curved surface on a flat surface? Keep reading to find out.

A Flat Sphere?

A map is a flat representation of the Earth's curved surface. However, when you move information from a curved surface to a flat surface, you lose some accuracy. Changes called *distortions* happen in the shapes and sizes of landmasses and oceans on maps. Direction and distance can also be distorted. Consider the example of the orange peel shown in **Figure 1**.

✓ **Reading Check** What are distortions on maps? (*See the Appendix for answers to Reading Checks.*)

Figure 1 *If you remove and flatten the peel from an orange, the peel will stretch and tear. Notice how shapes as well as distances between points on the peel are distorted.*

CHAPTER RESOURCES

Chapter Resource File

- **Lesson Plan**
- **Directed Reading A** BASIC
- **Directed Reading B** SPECIAL NEEDS

Technology

Transparencies
- Bellringer
- Cylindrical Projection; Conic Projection; Azimuthal Projection

Answer to Reading Check

Distortions are inaccuracies produced when information is transferred from a curved surface to a flat surface.

Map Projections

Mapmakers use map projections to move the image of Earth's curved surface onto a flat surface. No map projection of Earth can show the surface of a sphere in the correct proportions. All flat maps have distortion. However, a map showing a smaller area, such as a city, has less distortion than a map showing a larger area, such as the world.

To understand how map projections are made, think of Earth as a translucent globe that has a light inside. If you hold a piece of paper against the globe, shadows appear on the paper. These shadows show marks on the globe, such as continents, oceans, and lines of latitude and longitude. The way the paper is held against the globe determines the kind of map projection that is made. The most common map projections are based on three shapes—cylinders, cones, and planes.

Cylindrical Projection

A map projection that is made when the contents of the globe are moved onto a cylinder of paper is called a **cylindrical projection** (suh LIN dri kuhl proh JEK shuhn). The most common cylindrical projection is called a *Mercator projection* (muhr KAYT uhr proh JEK shuhn). The Mercator projection shows the globe's latitude and longitude lines as straight lines. Equal amounts of space are used between longitude lines. Latitude lines are spaced farther apart north and south of the equator. Because of the spacing, areas near the poles look wider and longer on the map than they look on the globe. In **Figure 2,** Greenland appears almost as large as Africa!

cylindrical projection a map projection that is made by moving the surface features of the globe onto a cylinder

INTERNET ACTIVITY

For another activity related to this chapter, go to **go.hrw.com** and type in the keyword **HZ5MAPW.**

Figure 2 · Cylindrical Projection

This cylindrical projection is a Mercator projection. It is accurate near the equator but distorts areas near the North and South Poles.

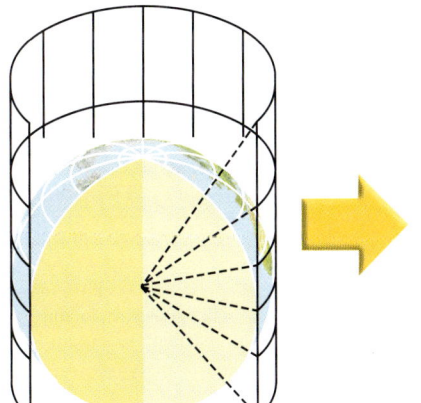

INTERNET ACTIVITY
Short Story — GENERAL

For an internet activity related to this chapter, have students go to **go.hrw.com** and type in the keyword **HZ5MAPW.**

SCIENCE HUMOR

Q: What do you get when you cross a cowboy with a mapmaker?

A: a cowtographer

READING STRATEGY — GENERAL

Mnemonics Have students think of some rhymes to help them remember key points about the projections discussed in the text. You might suggest the following to help students get started:

• "If you're traveling to the equator, you'll do well with Mercator."

• "For east to west, conic is best."

• "For a stroll at a pole, an azimuthal will help you stay in control." English Language Learners

LS Verbal/Auditory

Homework — GENERAL

Map Projections Have students make a chart listing the strengths and weaknesses of cylindrical, conic, azimuthal, and equal-area projections. Then, have students research another projection, such as the Robinson projection. Have them add the strengths and weaknesses of that projection to their charts. Have them explain at the bottom of the chart why none of the projections is entirely free of distortions and inaccuracies. **LS** Verbal

---

Figure 3 **Conic Projection**

A series of conic projections can be used to map a large area. Because each cone touches the globe at a different latitude, conic projections reduce distortion.

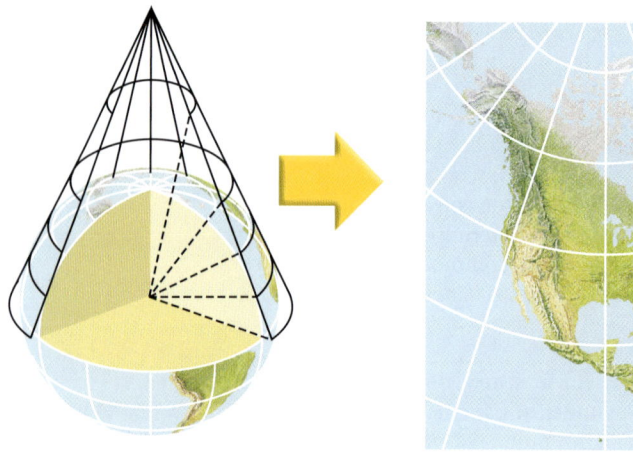

conic projection a map projection that is made by moving the surface features of the globe onto a cone

Conic Projection

A map projection that is made by moving the contents of the globe onto a cone is a **conic projection,** shown in **Figure 3.** This cone is then unrolled to form a flat plane.

The cone touches the globe at each line of longitude but at only one line of latitude. There is no distortion along the line of latitude where the globe touches the cone. Areas near this line of latitude are distorted less than other areas are. Because the cone touches many lines of longitude and only one line of latitude, conic projections are best for mapping large masses of land that have more area east to west. For example, a conic projection is often used to map the United States.

CONNECTION TO Social Studies

WRITING SKILL **Mapmaking and Ship Navigation** Gerardus Mercator is the cartographer (or mapmaker) who developed the Mercator projection. During his career as a mathematician and cartographer, Mercator worked hard to produce maps of many parts of Europe, including Great Britain. He also produced a terrestrial globe and a celestial globe. Use the library or the Internet to research Mercator. How did his mapmaking skills help ship navigators in the 1500s? Write a paragraph describing what you learn.

Science Bloopers

During the process of compiling data from a number of different sources, mistakes are sometimes made. Cartographers have wiped entire cities off maps accidentally! For example, Canada's capital, Ottawa, was once omitted from a Canadian tourist-office map. An official explanation was that there was no direct air service between New York City and Ottawa failed to satisfy one Ottawa tourist bureau executive, who remarked irately, "Ottawa should be shown in any case, even if the only point of entry was by two-man kayak."

Figure 4   **Azimuthal Projection**

On this azimuthal projection, distortion increases as you move farther from the North Pole.

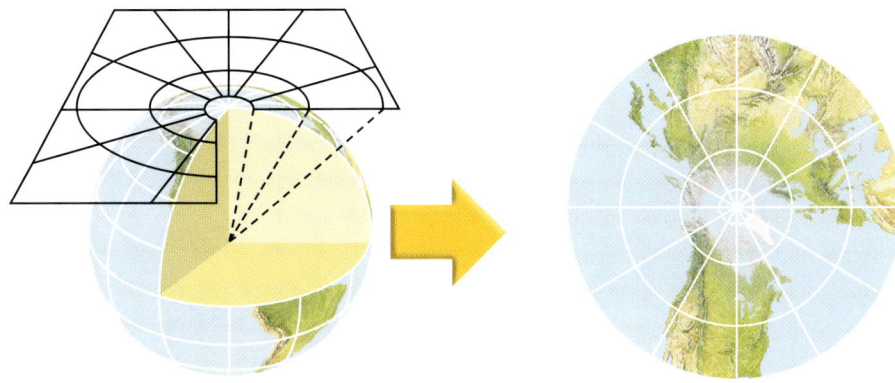

Azimuthal Projection

An **azimuthal projection** (AZ uh MYOOTH uhl proh JEK shuhn) is a map projection that is made by moving the contents of the globe onto a flat plane. Look at **Figure 4.** On an azimuthal projection, the plane touches the globe at only one point. There is little distortion at this point of contact. The point of contact for an azimuthal projection is usually one of the poles. However, distortion of direction, distance, and shape increases as you move away from the point of contact. Azimuthal projections are most often used to map areas of the globe that are near the North and South Poles.

azimuthal projection a map projection that is made by moving the surface features of the globe onto a plane

✔ **Reading Check**   How are azimuthal and conic projections alike? How are they different?

Equal-Area Projection

A map projection that shows the area between the latitude and longitude lines the same size as that area on a globe is called an *equal-area projection*. Equal-area projections can be made by using cylindrical, conic, or azimuthal projections. Equal-area projections are often used to map large land areas, such as continents. The shapes of the continents and oceans are distorted on equal-area projections. But because the scale used on equal-area projections is constant throughout the map, this type of projection is good for determining distance on a map. **Figure 5** is an example of an equal-area projection.

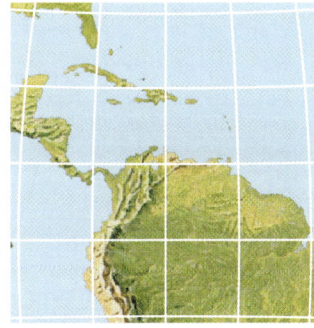

Figure 5 *Equal-area projections are useful for determining distance on a map.*

• Learning Disabled
• Attention Deficit Disorder
• Gifted and Talented

Organize students into pairs or groups of three. Hand out to each group a map of their town or city and brightly colored sticky dots. Ask students to identify and mark with sticky dots on the map the title of the map, map scale, compass rose, and legend. Ask them to find their school, home, and an additional landmark and indicate these sites as well. Next, ask each team to calculate the distance from their school to the various landmarks. For gifted and talented students, ask them to do additional calculations, such as distance between points and direction in degrees using a compass.

English Language Learners

LS Visual

Answer to Reading Check
Every map should have a title, a compass rose, a scale, the date, and a legend.

Information Shown on Maps

Regardless of the kind of map you are reading, the map should contain the information shown in **Figure 6.** This information includes a title, a compass rose, a scale, a legend, and a date. Unfortunately, not all maps have all this information. The more of this information a map has, the more reliable the map is.

Reading Check What information should every map have?

Figure 6 This Texas road map includes all of the information that a map should contain.

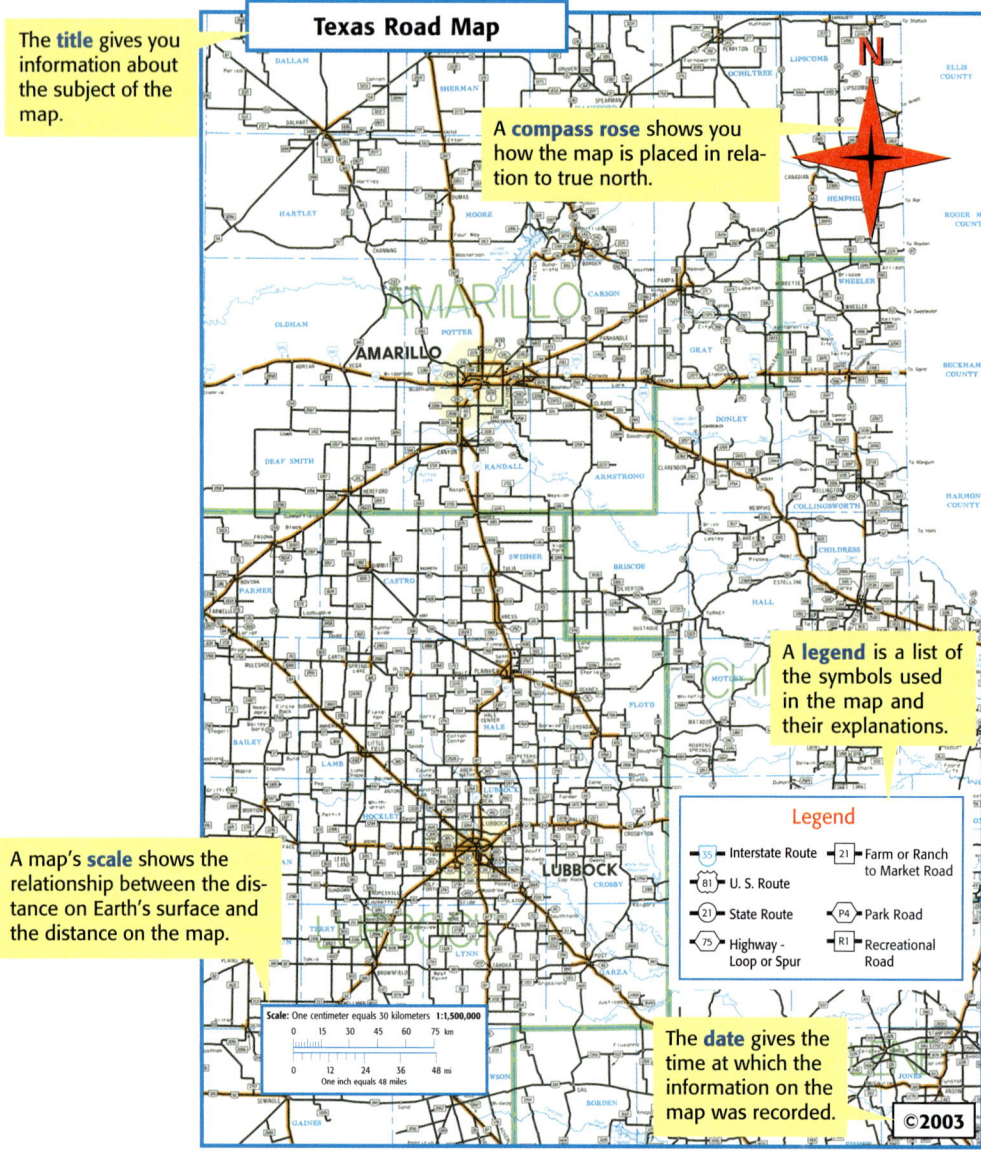

Texas Road Map

The **title** gives you information about the subject of the map.

A **compass rose** shows you how the map is placed in relation to true north.

A **legend** is a list of the symbols used in the map and their explanations.

A map's **scale** shows the relationship between the distance on Earth's surface and the distance on the map.

Legend

35 Interstate Route	21 Farm or Ranch to Market Road
81 U. S. Route	
21 State Route	P4 Park Road
75 Highway - Loop or Spur	R1 Recreational Road

Scale: One centimeter equals 30 kilometers 1:1,500,000
0   15   30   45   60   75 km
0   12   24   36   48 mi
One inch equals 48 miles

The **date** gives the time at which the information on the map was recorded.

©2003

CONNECTION ACTIVITY
Math ——— GENERAL

Taking a Hike Have students imagine that they want to use a map with a scale of 1:24,000 to estimate the length of a hike. On the map, the route measures 20 cm. Have students calculate the length of the hike in kilometers.
(20 cm × 24,000 = 480,000 cm;
480,000 cm ÷ 100 cm/m = 4,800 m;
4,800 ÷ 1,000 m/km = 4.8 km)
LS Logical

Is That a Fact!

In an attempt to keep foreigners and even Soviet citizens from knowing the exact geography of their country, mapmakers in the Soviet Union printed maps with deliberate mistakes. Roads, rivers, and even cities would be misplaced or omitted. These revisions were especially true of Moscow street maps. Although these maps were made with national security in mind, the end result was that the country operated much less efficiently.

Modern Mapmaking

For many centuries mapmakers relied on the observations of explorers to make maps. Today, however, mapmakers have far more technologically advanced tools for mapmaking.

Many of today's maps are made by remote sensing. **Remote sensing** is a way to collect information about something without physically being there. Remote sensing can be as basic as putting cameras on airplanes. However, many mapmakers rely on more sophisticated technology, such as satellites.

Remote Sensing and Satellites

The image shown in **Figure 7** is a photograph taken by a satellite. Satellites can also detect energy that your eyes cannot. Remote sensors gather data about energy coming from Earth's surface and send the data back to receiving stations on Earth. A computer is then used to process the information to make a picture you can see.

Remote Sensing Using Radar

Radar is a tool that uses waves of energy to map Earth's surface. Waves of energy are sent from a satellite to the area being observed. The waves are then reflected from the area to a receiver on the satellite. The distance and the speed in which the waves travel to the area and back are measured and analyzed to create a map of the area. The waves used in radar can move through clouds and water. Because of this ability, radar has been used to map the surface of Venus, whose atmosphere is thick and cloudy.

remote sensing the process of gathering and analyzing information about an object without physically being in touch with the object

Figure 7 *Satellites can produce very detailed images of the Earth's surface. The satellite that took this picture was 423 mi above the Earth's surface!*

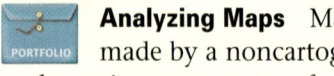

Figure 8 This tiny GPS unit may come in handy if you are ever lost.

Global Positioning System

Did you know that satellite technology can actually help you from getting lost? The *global positioning system* (GPS) can help you find where you are on Earth. GPS is a system of orbiting satellites that send radio signals to receivers on Earth. The receivers calculate a given place's latitude, longitude, and elevation.

GPS was invented in the 1970s by the U.S. Department of Defense for military use. However, during the last 30 years, GPS has made its way into people's daily lives. Mapmakers use GPS to verify the location of boundary lines between countries and states. Airplane and boat pilots use GPS for navigation. Businesses and state agencies use GPS for mapping and environmental planning. Many new cars have GPS units that show information on a screen on the dashboard. Some GPS units are small enough to wear on your wrist, as shown in **Figure 8,** so you can know your location anywhere you go!

Geographic Information Systems

Mapmakers now use geographic information systems to store, use, and view geographic information. A *geographic information system*, or GIS, is a computerized system that allows a user to enter different types of information about an area. This information is entered and stored as layers. The user can then use the stored information to make complex analyses or display maps. **Figure 9** shows three GIS images of Seattle, Washington.

✓ Reading Check Explain how information is stored using GIS.

Figure 9 The images at right show the location of sewer lines, roads, and parks in Seattle, Washington.

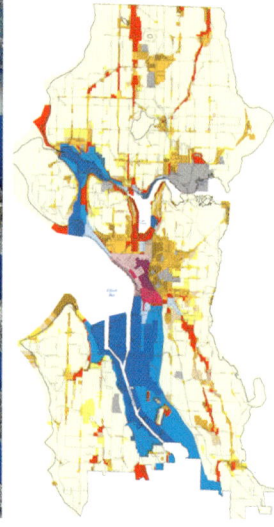

Answer to Reading Check
A GIS system stores information in layers.

Summary

- When information is moved from a curved surface to a flat surface, distortion occurs.
- Three main types of projections are used to show Earth's surface on a flat map: cylindrical, conic, and azimuthal projections.
- Equal-area maps are used to show the area of a piece of land in relation to the area of other landmasses and oceans.
- Maps should contain a title, a scale, a legend, a compass rose, and a date.

- Modern mapmakers use remote sensing technology, such as satellites and radar.
- The Global positioning system, or GPS, is a system of satellites that can help you determine your location no matter where you are.
- Geographical information systems, or GIS, are computerized systems that allow mapmakers to store and use many types of data about an area.

Using Key Terms

1. In your own words, write a definition for each of the following terms: *cylindrical projection, azimuthal projection,* and *conic projection.*

Understanding Key Ideas

2. Which of the following map projections is most often used to map the United States?
   a. cylindrical projection
   b. conic projection
   c. azimuthal projection
   d. equal-area projection

3. List five things found on maps. Explain how each thing is important to reading a map.

4. Describe how GPS can help you find your location on Earth.

5. Why is radar useful when mapping areas that tend to be covered in clouds?

Critical Thinking

6. **Analyzing Ideas** Imagine you are a mapmaker. You have been asked to map a landmass that has more area from east to west than from north to south. What type of map projection would you use? Explain.

7. **Making Inferences** Imagine looking at a map of North America. Would this map have a large scale or a small scale? Would a map of your city have a large scale or a small scale? Explain.

Interpreting Graphics

Use the map below to answer the questions that follow.

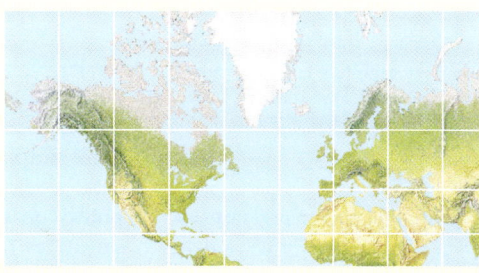

8. What type of projection was used to make this map?

9. Which areas of this map are the most distorted? Explain.

10. Which areas of this map are the least distorted? Explain.

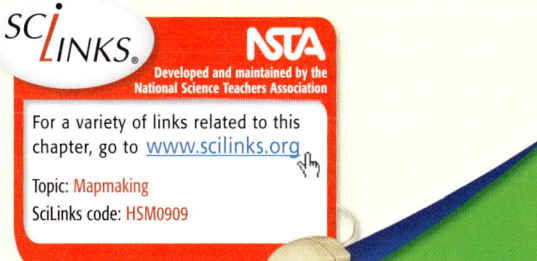

Developed and maintained by the National Science Teachers Association

For a variety of links related to this chapter, go to www.scilinks.org

Topic: Mapmaking
SciLinks code: HSM0909

SECTION
3

Focus

Overview

In this section, students investigate how contour lines are used to show elevation and landforms on a topographic map. In addition, they learn how to read and interpret the features of a topographic map.

🔔 Bellringer

Have students examine the topographic map shown on this page. Have them imagine that they are standing on the top of Campbell Hill. Students should describe in their **science journal** what they see in each direction. Tell students that they will learn to read topographic maps, such as the ones in this section.

Motivate

ACTIVITY ———— GENERAL

Investigate Your Area If possible, obtain topographic maps of your area. You may be able to find these maps at camping stores or on the Internet. As a class, locate different landforms in your area. Discuss with students how contour intervals indicate changes in elevation. If possible, take a class field trip to one of the areas. **LS Visual** English Language Learners

---

READING WARM-UP

Objectives

- Explain how contour lines show elevation and landforms on a map.
- Explain how the relief of an area determines the contour interval used on a map.
- List the rules of contour lines.

Terms to Learn

topographic map
elevation
contour line
contour interval
relief
index contour

READING STRATEGY

Paired Summarizing Read this section silently. In pairs, take turns summarizing the material. Stop to discuss ideas that seem confusing.

topographic map a map that shows the surface features of Earth

elevation the height of an object above sea level

contour line a line that connects points of equal elevation

Topographic Maps

Imagine you are going on a camping trip in the wilderness. To be prepared, you want to take a compass and a map. But what kind of map should you take? Because there won't be any roads in the wilderness, you can forget about a road map. Instead, you will need a topographic map.

A **topographic map** (TAHP uh GRAF ik MAP) is a map that shows surface features, or topography (tuh PAHG ruh fee), of the Earth. Topographic maps show both natural features, such as rivers, lakes, and mountains, and features made by humans, such as cities, roads, and bridges. Topographic maps also show elevation. **Elevation** is the height of an object above sea level. The elevation at sea level is 0. In this section, you will learn how to read a topographic map.

Elements of Elevation

The United States Geological Survey (USGS), a federal government agency, has made topographic maps for most of the United States. These maps show elevation in feet (ft) rather than in meters, the SI unit usually used by scientists.

Contour Lines

On a topographic map, *contour lines* are used to show elevation. **Contour lines** are lines that connect points of equal elevation. For example, one contour line would connect points on a map that have an elevation of 100 ft. Another line would connect points on a map that have an elevation of 200 ft. **Figure 1** illustrates how contour lines appear on a map.

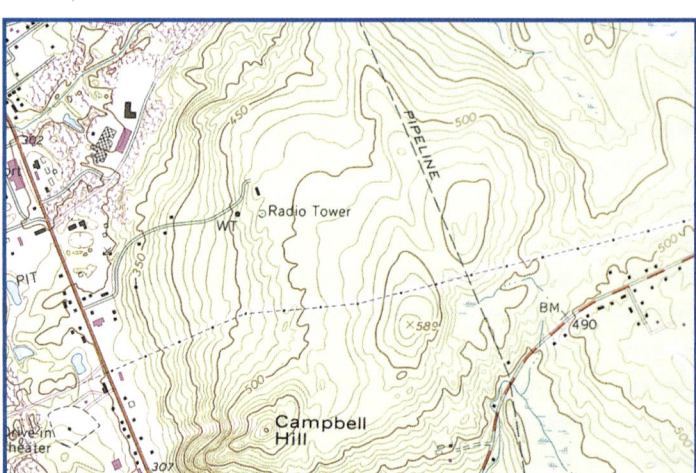

Figure 1 *Because contour lines connect points of equal elevation, the shape of the contour lines reflects the shape of the land.*

CHAPTER RESOURCES

Chapter Resource File

- Lesson Plan
- Directed Reading A **BASIC**
- Directed Reading B **SPECIAL NEEDS**

Technology

Transparencies
- Bellringer

Is That a Fact!

The Ordnance Survey of Great Britain produces topographic maps with very large scales, ranging from 1:10,000 to 1:1,250. Such large scales permit a level of detail that shows the location of public telephones, windmills, and even large boulders!

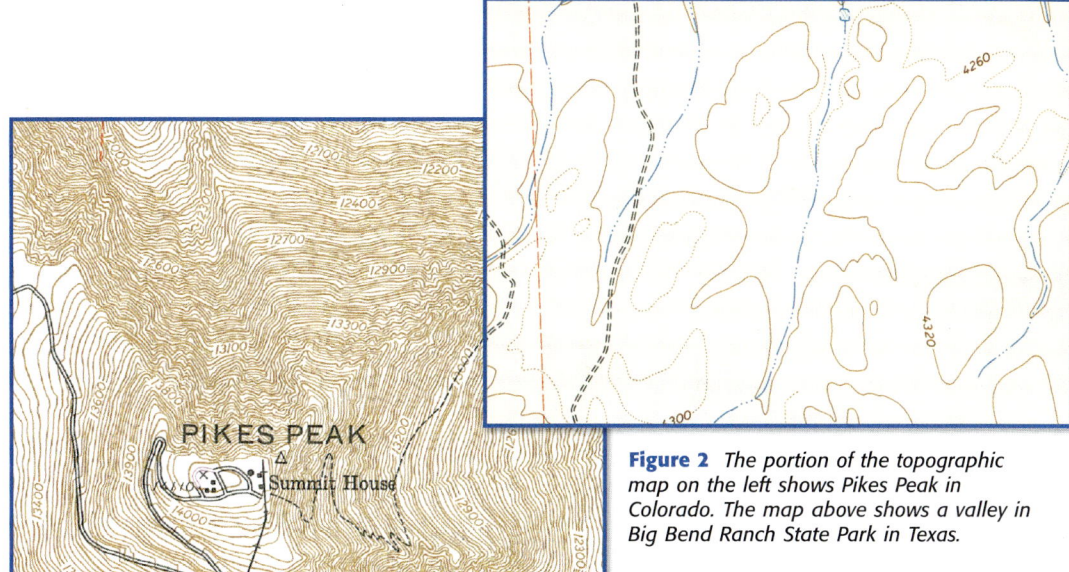

Figure 2 *The portion of the topographic map on the left shows Pikes Peak in Colorado. The map above shows a valley in Big Bend Ranch State Park in Texas.*

Contour Interval

The difference in elevation between one contour line and the next is called the **contour interval.** For example, a map with a contour interval of 20 ft would have contour lines every 20 ft of elevation change, such as 0 ft, 20 ft, 40 ft, and 60 ft. A mapmaker chooses a contour interval based on the area's relief. **Relief** is the difference in elevation between the highest and lowest points of the area being mapped. Because the relief of an area with mountains is large, the relief might be shown on a map using a large contour interval, such as 100 ft. However, a flat area has small relief and might be shown on a map by using a small contour interval, such as 10 ft.

The spacing of contour lines also indicates slope, as shown in **Figure 2.** Contour lines that are close together show a steep slope. Contour lines that are spaced far apart show a gentle slope.

Index Contour

On USGS topographic maps, an index contour is used to make reading the map easier. An **index contour** is a darker, heavier contour line that is usually every fifth line and that is labeled by elevation. Find an index contour on both of the topographic maps shown in **Figure 2.**

✓ *Reading Check* What is an index contour? (*See the Appendix for answers to Reading Checks.*)

contour interval the difference in elevation between one contour line and the next

relief the variations in elevation of a land surface

index contour on a map, a darker, heavier contour line that is usually every fifth line and that indicates a change in elevation

CONNECTION TO Oceanography

Mapping the Ocean Floor Oceanographers use topographic maps to map the topography of the ocean floor. Use the Internet or the library to find a topographic map of the ocean floor. How are maps of the ocean floor similar to maps of the continents? How are they different?

Rules of Contour Lines Using a topographic map, show students examples of the rules of contour lines. Discuss how contour lines never cross, and show specific areas of steep slope, gentle slope, rivers or streams, and hills and depressions. **LS Visual**

Quiz — GENERAL

1. What is a contour interval on a topographic map? (the difference in elevation between one contour line and the next)

2. What do closely spaced contour lines on a topographic map indicate? (a steep area)

3. How does a topographic map indicate the direction that a stream flows? (Streams flow downhill, or in the direction that elevation decreases. The Vs point toward higher elevations.)

Alternative Assessment — GENERAL

Using Maps Photocopy a portion of a topographic map that shows a mountain. Indicate the scale of the map on the photocopy. Distribute copies to students, and ask them to trace a route to the top. Then, have them write a description of their "trail" that includes the length of the hike, the elevations where the trail is the steepest, and the points where the slope is gentle. Have students also note other features, such as streams, power lines, or road crossings. **LS Verbal**

Reading a Topographic Map

Topographic maps, like other maps, use symbols to represent parts of the Earth's surface. **Figure 3** shows a USGS topographic map. The legend shows some of the symbols that represent features in topographic maps.

Colors are also used to represent features of Earth's surface. In general, buildings, roads, bridges, and railroads are black. Contour lines are brown. Major highways are red. Bodies of water, such as rivers, lakes, and oceans are blue. Cities and towns are pink, and wooded areas are green.

Figure 3 *All USGS topographic maps use the same symbols to show natural and human-made features.*

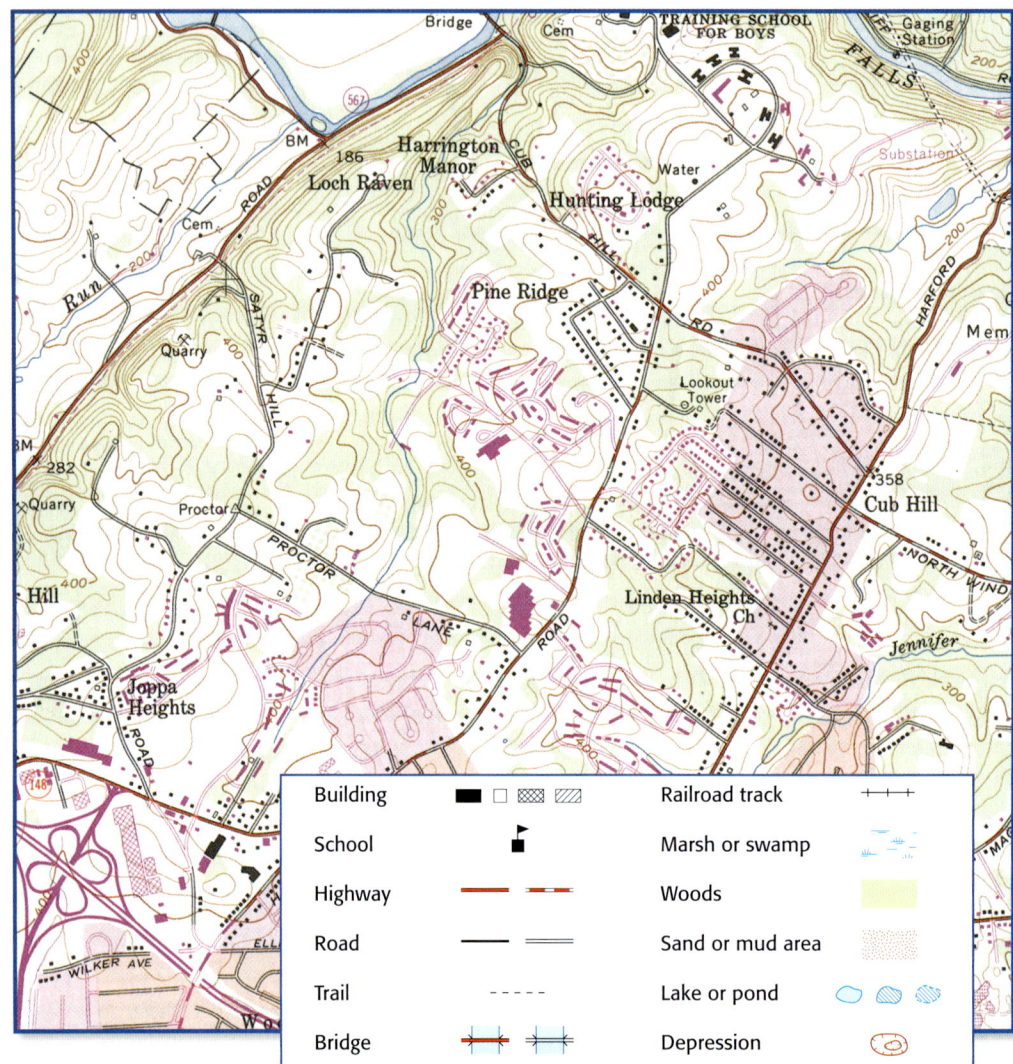

The Golden Rules of Contour Lines

Contour lines are the key to explaining the size and shape of landforms on a topographic map. Reading a topographic map takes training and practice. The following rules will help you understand how to read topographic maps:

- Contour lines never cross. All points along a contour line represent one elevation.

- The spacing of contour lines depends on slope characteristics. Contour lines that are close together show a steep slope. Contour lines that are far apart show a gentle slope.

- Contour lines that cross a valley or stream are V shaped. The V points toward the area of highest elevation. If a stream or river flows through the valley, the V points upstream.

- The tops of hills, mountains, and depressions are shown by closed circles. Depressions are marked with short, straight lines inside the circle that point downslope to the depression.

CONNECTION TO Environmental Science

Endangered Species State agencies, such as the Texas Parks and Wildlife Department, use topographic maps to mark where endangered plant and animal species are. By marking the location of the endangered plants and animals, these agencies can record and protect these places. Use the Internet or another source to find out if there is an agency in your state that tracks endangered species by using topographic maps.

SECTION Review

Summary

- Contour lines are used to show elevation and landforms by connecting points of equal elevation.

- The contour interval is determined by the relief of an area.

- Contour lines never cross. Contour lines that cross a valley or a stream are V shaped and point upstream. The tops of hills, mountains, and depressions are shown by closed circles.

Using Key Terms

1. In your own words, write a definition for each of the following terms: *topographic map, contour interval,* and *relief.*

Understanding Key Ideas

2. An index contour
   - **a.** is a heavier contour line that shows a change in elevation.
   - **b.** points in the direction of higher elevation.
   - **c.** indicates a depression.
   - **d.** indicates a hill.

3. How do topographic maps represent the Earth's surface?

4. How does the relief of an area determine the contour interval used on a map?

5. What are the rules of contour lines?

Math Skills

6. The contour line at the base of a hill reads 90 ft. There are five contour lines between the base of the hill and the top of the hill. If the contour interval is 30 ft, what is the elevation of the highest contour line?

Critical Thinking

7. **Making Inferences** Why isn't the highest point on a hill represented by a contour line?

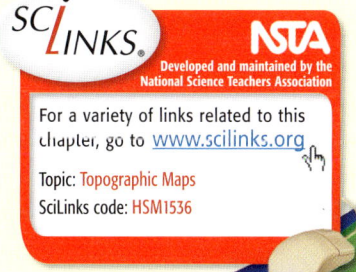

SCI LINKS.

NSTA
Developed and maintained by the National Science Teachers Association

For a variety of links related to this chapter, go to www.scilinks.org

Topic: Topographic Maps
SciLinks code: HSM1536

Answers to Section Review

1. Sample answer: A topographic map shows the surface features of the Earth. A contour interval is the difference in elevation between two adjacent contour lines. Relief is the difference in elevation between the highest and lowest points of the area being mapped.

2. a

3. Topographic maps use contour lines to show the surface features of Earth. Symbols indicate other features.

4. If the relief of an area is large, a topographic map of the area will have a large contour interval. If the relief of an area is small, a topographic map of the area will have a small contour interval.

5. Contour lines never cross. The spacing of contour lines depends on the slope characteristics of an area. Contour lines that cross a valley or a stream are V shaped, and the V points upstream. The tops of mountains and the bottoms of depressions are shown by closed circles. Depressions are marked with short, straight lines inside the circle that point downslope.

6. 30 ft × 5 = 150 ft; 90 ft + 150 ft = 240 ft

7. The highest point on a hill or mountain is a single point, not a group of points with the same elevation.

CHAPTER RESOURCES

Chapter Resource File

 • Section Quiz GENERAL
- Section Review GENERAL
- Vocabulary and Section Summary GENERAL
- Reinforcement Worksheet BASIC

Workbooks

 Math Skills for Science
- Mapping and Surveying GENERAL

Round or Flat?

Teacher's Notes

Time Required

One 45-minute class period

Lab Ratings

EASY ——————————— HARD

Teacher Prep 🧪🧪

Student Set-Up 🧪🧪🧪

Concept Level 🧪🧪🧪

Clean Up 🧪

MATERIALS

The materials listed on the student page are enough for a group of 3 or 4 students.

Safety Caution

Remind students to review all safety cautions and icons before beginning this lab activity.

Preparation Notes

Obtain inflated basketballs from your school's physical education instructor. Asking students to bring basketballs from home may be necessary. Begin the activity by reminding students that circumference is the distance around a circle or sphere.

You may also need to review the use of protractors with students before performing this activity.

OBJECTIVES

Construct a tool to measure the circumference of the Earth.

Calculate the circumference of the Earth.

MATERIALS

- basketball
- books or notebooks (2)
- calculator (optional)
- clay, modeling
- flashlight or small lamp
- meterstick
- pencils, unsharpened (2)
- protractor
- ruler, metric
- string, 10 cm long
- tape, masking
- tape measure

SAFETY

Round or Flat?

Eratosthenes thought of a way to measure the circumference of Earth. He came up with the idea when he read that a well in southern Egypt was entirely lit by the sun at noon once each year. He realized that to shine on the entire surface of the well water, the sun must be directly over the well. At the same time, in a city just north of the well, a tall monument cast a shadow. Thus, Eratosthenes reasoned that the sun could not be directly over both the monument and the well at noon on the same day. In this experiment, you will see how Eratosthenes' way of measuring works.

Ask a Question

❶ How could I use Eratosthenes' method of investigation to measure the size of the Earth?

Form a Hypothesis

❷ Formulate a hypothesis that answers the question above. Record your hypothesis.

Test the Hypothesis

❸ Set the basketball on a table. Place a book or notebook on either side of the basketball to hold the ball in place. The ball represents Earth.

Lab Notes

Explain that Eratosthenes' experiment worked because he set up a ratio. It may be necessary to review ratios before performing this activity. The formula Eratosthenes used is as follows:

$$\frac{distance\ around\ ball}{distance\ between\ sticks} = \frac{360°\ in\ the\ sphere}{angle\ of\ shadow\ with\ stick}$$

Students may be interested to learn that they can calculate the circumference of the Earth by performing Eratosthenes' experiment in partnership with other schools around the world. The experiment is conducted twice a year during the fall and spring equinoxes. To find out more, have students search for "Eratosthenes' experiment" on the Internet.

4 Use modeling clay to attach a pencil to the "equator" of the ball so that the pencil points away from the ball.

5 Attach the second pencil to the ball at a point that is 5 cm above the first pencil. This second pencil should also point away from the ball.

6 Use a meterstick to measure 1 m away from the ball. Mark the 1 m position with masking tape. Label the position "Sun." Hold the flashlight so that its front edge is above the masking tape.

7 When your teacher turns out the lights, turn on your flashlight and point it so that the pencil on the equator does not cast a shadow. Ask a partner to hold the flashlight in this position. The second pencil should cast a shadow on the ball.

8 Tape one end of the string to the top of the second pencil. Hold the other end of the string against the ball at the far edge of the shadow. Make sure that the string is tight. But be careful not to pull the pencil over.

9 Use a protractor to measure the angle between the string and the pencil. Record this angle.

10 Use the following formula to calculate the experimental circumference of the ball.

$$\text{Circumference} = \frac{360° \times 5 \text{ cm}}{\text{angle between pencil and string}}$$

11 Record the experimental circumference you calculated in step 10. Wrap the tape measure around the ball's equator to measure the actual circumference of the ball. Record this circumference.

Analyze the Results

1 **Examining Data** Compare the experimental circumference with the actual circumference.

2 **Analyzing Data** What could have caused your experimental circumference to differ from the actual circumference?

3 **Analyzing Data** What are some of the advantages and disadvantages of taking measurements this way?

Draw Conclusions

4 **Evaluating Methods** Was Eratosthenes' method an effective way to measure Earth's circumference? Explain your answer.

Barry L. Bishop
San Rafael Junior High
Ferron, Utah

Chapter Review

Assignment Guide

SECTION	QUESTIONS
1	1–3, 7, 8
2	4, 6, 10, 11, 12, 14–16, 18–22, 24
3	5, 9, 13, 17, 23, 25–29

ANSWERS

Using Key Terms

1. Sample answer: True north is the geographic North Pole. Magnetic north refers to the magnetic north pole, which changes.

2. Sample answer: Latitude is the distance north or south from the equator. Longitude is the distance east and west from the prime meridian. Both latitude and longitude are measured in degrees.

3. Sample answer: The equator is the imaginary circle halfway between the poles that divides the Earth into Northern and Southern Hemispheres and represents 0° latitude. The prime meridian represents 0° longitude. It runs from the North to South Poles through Greenwich, England.

USING KEY TERMS

For each pair of terms, explain how the meanings of the terms differ.

1. *true north* and *magnetic north*

2. *latitude* and *longitude*

3. *equator* and *prime meridian*

4. *cylindrical projection* and *azimuthal projection*

5. *contour interval* and *index contour*

6. *global positioning system* and *geographic information system*

UNDERSTANDING KEY IDEAS

Multiple Choice

7. A point whose latitude is 0° is located on the
   a. North Pole.
   b. equator.
   c. South Pole.
   d. prime meridian.

8. The distance in degrees east or west of the prime meridian is
   a. latitude.
   b. declination.
   c. longitude.
   d. projection.

9. Widely spaced contour lines indicate a
   a. steep slope.
   b. gentle slope.
   c. hill.
   d. river.

10. The most common map projections are based on three geometric shapes. Which of the following geometric shapes is NOT one of the three geometric shapes?
   a. cylinder
   b. square
   c. cone
   d. plane

11. A cylindrical projection is distorted near the
   a. equator.
   b. poles.
   c. prime meridian.
   d. date line.

12. What is the relationship between the distance on a map and the actual distance on Earth called?
   a. legend
   b. elevation
   c. relief
   d. scale

13. ___ is the height of an object above sea level.
   a. Contour interval
   b. Elevation
   c. Declination
   d. Index contour

Short Answer

14. List four methods that modern mapmakers use to make accurate maps.

15. Why is a map legend important?

4. Sample answer: A cylindrical projection is a map projection made by transferring the surface of the globe onto a cylinder. An azimuthal projection is a map projection made by projecting the surface of the globe onto a plane.

5. Sample answer: Contour interval is the difference in elevation between one contour line and the next. An index contour is a darker, heavier contour line that usually occurs every fifth line.

6. Sample answer: The global positioning system is a system of satellites that send radio signals to receivers on Earth. The receivers calculate a given place's latitude, longitude, and elevation. A geographic information system is a computer system that stores data about an area in layers.

Understanding Key Ideas

7. b	11. b
8. c	12. d
9. b	13. b
10. b	

16. Why does Greenland appear so large in relation to other landmasses on a map made using a cylindrical projection?

17. What is the function of contour lines on a topographic map?

18. How can GPS help you find your location on Earth?

19. What is GIS?

CRITICAL THINKING

20. Concept Mapping Use the following terms to create a concept map: *maps, legend, map projection, map parts, scale, cylinder, title, cone, plane, date,* and *compass rose.*

21. Making Inferences One of the important parts of a map is its date. Why is the date important?

22. Analyzing Ideas Why is it important for maps to have scales?

23. Applying Concepts Imagine that you are looking at a topographic map of the Grand Canyon. Would the contour lines be spaced close together or far apart? Explain your answer.

24. Analyzing Processes How would a GIS system help a team of engineers plan a new highway system for a city?

25. Making Inferences If you were stranded in a national park, what kind of map of the park would you want to have with you? Explain your answer.

INTERPRETING GRAPHICS

Use the topographic map below to answer the questions that follow.

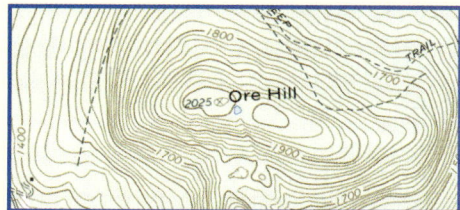

26. What is the elevation change between two adjacent lines on this map?

27. What type of relief does this area have?

28. What surface features are shown on this map?

29. What is the elevation at the top of Ore Hill?

14. Modern mapmakers use satellite and radar technology, GPS, and GIS to make accurate maps.

15. A map legend is important because it defines the set of symbols used in the map.

16. Greenland appears large on a map created with a cylindrical projection because of distortion. Maps created with a cylindrical projection are increasingly distorted as the distance from the equator increases.

17. Contour lines on a topographic map show the elevation, the relief, and the shape of landforms.

CHAPTER RESOURCES

Chapter Resource File
- Chapter Review **GENERAL**
- Chapter Test A **GENERAL**
- Chapter Test B **ADVANCED**
- Chapter Test C **SPECIAL NEEDS**
- Vocabulary Activity **GENERAL**

Workbooks

Study Guide
- Assessment resources are also available in Spanish.

18. A GPS unit receives signals from a system of satellites and then calculates the latitude and longitude of your location.

19. GIS is a computerized system that allows a user to enter different types of information about an area, this information is then stored in layers.

Critical Thinking

20. 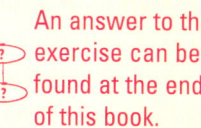 An answer to this exercise can be found at the end of this book.

21. Sample answer: A date on a map is important because the Earth's surface is constantly changing. The date shows you how old the information is.

22. Sample answer: It is important that maps have scales because a map scale allows the user to determine distances on the map.

23. Sample answer: A topographic map of the Grand Canyon would show contour lines close together, indicating steep slopes.

24. Sample answer: A GIS system would allow the team of engineers to access information about current roads and streets, as well as power lines, sewer lines, natural features, and any other information they may need when building a highway.

25. Sample answer: I would want to have a topographic map because this type of map shows the relief of the area rather than only roads and streets.

Interpreting Graphics

26. 20 ft

27. It has very large relief.

28. Answers may vary. Sample answer: Two hills are shown on this map.

29. 2,025 ft

Standardized Test Preparation

Teacher's Note

To provide practice under more realistic testing conditions, give students 20 minutes to answer all of the questions in this Standardized Test Preparation.

MISCONCEPTION
ALERT

Answers to the standardized test preparation can help you identify student misconceptions and misunderstandings.

READING

Passage 1

1. C

2. F

3. C

 **TEST DOCTOR**

Question 1: Students may think that the last sentence means that maps can show accurate sizes of Earth's oceans as well as show a lot of detail. However, the passage states in the second sentence that globes and maps are both models of Earth.

Question 3: Some students may misunderstand the idea that globes show less detail than maps. Therefore, some students may choose one of the answer choices concerning details found mainly on maps rather than on globes.

READING

Read each of the passages below. Then, answer the questions that follow each passage.

Passage 1 Scientists use models to represent things. Globes and maps are examples of models that scientists use to study Earth's surface.

Because a globe is a sphere, as Earth is, a globe is the most accurate model of Earth. A globe accurately shows the sizes and shapes of the continents and oceans in relation to one another. But a globe is not always the best model to use when studying Earth's surface. For example, a globe is too small to show a lot of detail, such as roads and rivers. It is much easier to show details on maps. Maps can show the whole Earth or parts of it.

1. According to the passage, how are a globe and a map alike?
- **A** Both show a lot of detail.
- **B** Both are used to study the size of Earth's oceans.
- **C** Both are models used to the study Earth's surface.
- **D** Both show one part of Earth.

2. How are a globe and Earth alike?
- **F** Both are spheres.
- **G** Both represent real things.
- **H** Both are flat surfaces.
- **I** Both are models.

3. According to the passage, examining a globe would help you answer which of the following questions?
- **A** How many highways are in Michigan?
- **B** Where are the streams in my state?
- **C** Which continents border the Indian Ocean?
- **D** What is the exact length of the Nile River?

Passage 2 The names of many geographic locations in the United States are <u>rich</u> in description and national history. Names such as Adirondack and Chesapeake come from Native American languages. Some names, such as New London, Baton Rouge, and San Francisco, reflect European naming traditions. Other names, such as Stone Mountain and Long Island, provide a description of the area. The mapping efforts in the United States that took place after the Civil War often led to multiple names for one location. But mapmakers and scientists needed consistent names of locations for their studies. In 1890, the U.S. Board on Geographic Names was formed. This board determines and maintains location names.

1. In the passage, what does *rich* mean?
- **A** wealthy
- **B** abundant
- **C** incomplete
- **D** thick

2. Which of the following statements is true?
- **F** Mapmakers enjoyed using multiple names for the same location.
- **G** The U.S. Board on Geographic Names determines the name for an area.
- **H** Names such as Baton Rouge and San Francisco describe the physical area.
- **I** All geographic names came from Native American languages.

3. What can you infer from the passage?
- **A** Mapmakers name locations after themselves.
- **B** Scientists used descriptions of the physical area as names for locations.
- **C** The U.S. Board on Geographic Names now determines the names for locations.
- **D** Today, many locations in the United States have several names.

Passage 2

1. B

2. G

3. C

 **TEST DOCTOR**

Question 1: Some students may interpret the meaning of the word rich to mean "wealthy" in this context. Therefore, students may choose answer A. However, in this passage, the meaning of the word rich is "abundant."

Use each figure below to answer the question that follows each figure.

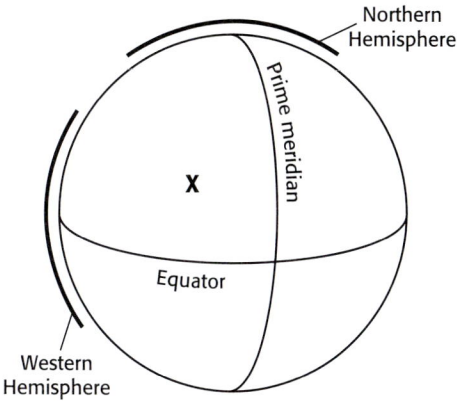

Northern Hemisphere

Prime meridian

X

Equator

Western Hemisphere

1. An X shows the location of a field investigation study site. Which of the following pairs of terms accurately describes the location of the X on the map?

A Northern Hemisphere; Western Hemisphere
B Northern Hemisphere; Eastern Hemisphere
C Southern Hemisphere; Eastern Hemisphere
D Southern Hemisphere; Western Hemisphere

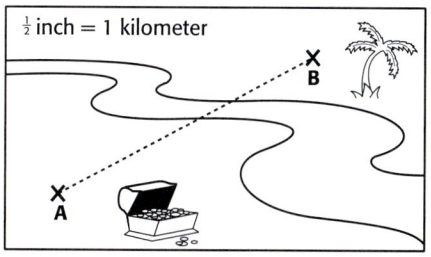

$\frac{1}{2}$ inch = 1 kilometer

X B

X A

2. The map above shows the distance from point A to point B. According to this map, what is the actual distance from point A to point B?

F 1 km
G 2 km
H 4 km
I 6 km

Read each question below, and choose the best answer.

1. Greenland's area is approximately 2 million square kilometers. The area of Africa is approximately 15 times the area of Greenland. What is the approximate area of Africa?

A 30 million square kilometers
B 17 million square kilometers
C 13 million square kilometers
D 7.5 million square kilometers

2. A satellite is 264 km above Earth's surface. What is this measurement expressed in meters?

F 264,000 m
G 26,400 m
H 2,640 m
I 0.264 m

3. On a topographic map, every fifth contour line is a darker line, or *index contour*. How many index contours are there in a series of 50 contour lines?

A 8
B 9
C 10
D 11

4. Juan and Maria hike up a mountain. Maria is at an elevation of 4.3 km. Juan is at an elevation of 2.7 km. What is the difference between their elevations?

F 1.6 km
G 2.6 km
H 6.0 km
I 7.0 km

5. The North Pole is 90°N latitude. If you drew a line from the North Pole to the center of Earth and a line from a point on the equator to the center of Earth, what kind of angle would the two lines form at Earth's center?

A acute
B obtuse
C equilateral
D right

Standardized Test Preparation

1. A
2. H

✚ TEST DOCTOR

Question 2: Students must be able to estimate the length of an inch in order to answer this question. Some students may not estimate this distance correctly.

1. A
2. F
3. C
4. F
5. D

✚ TEST DOCTOR

Question 1: Students may mistakenly add or subtract the estimated areas of Greenland and Africa. Students must multiply the estimated areas of Greenland and Africa in order to arrive at the correct answer.

Question 5: Students must have some prior knowledge of basic geometry in order to recognize that a 90° angle is a right angle. In addition, some students may not recognize the equator as being 0° latitude and therefore may answer incorrectly.

CHAPTER RESOURCES

Chapter Resource File

• Standardized Test Preparation GENERAL

State Resources

For specific resources for your state, visit **go.hrw.com** and type in the keyword **HSMSTR**.

Science, Technology, and Society

ACTIVITY —————— GENERAL

If you have access to a GPS unit, try taking your class on a geo-cache treasure hunt. You can find geocaches posted at many different sites on the Internet. Simply enter in the keyword *geocaching*. Choose a geocache that is relatively easy to find. You may want to find the cache first before taking your class with you to ensure that the cache is still there and that your students can safely access the area.

Scientific Discoveries

Background

According to legend, Allah became displeased with the wickedness of the citizens of Ubar and buried the city under a wave of sand. Ubar was lost for millennia until filmmaker Nicholas Clapp, NASA scientist Dr. Ronald Blom, and a team of explorers uncovered the ruins in 1991.

Discussion —————— GENERAL

Discuss with students the myth of the lost city of Atlantis. Have students suggest different ways that archeologists might search for a city that was covered by water instead of by sand.

Science in Action

Scientific Discoveries

The Lost City of Ubar

According to legend, the city of Ubar was a prosperous ancient city. Ubar was most famous for its frankincense, a tree sap that had many uses. As Ubar was in its decline, however, something strange happened. The city disappeared! It was a great myth that Ubar was swallowed up by the desert. It wasn't until present-day scientists used information from a Shuttle Imaging Radar system aboard the space shuttle that this lost city was found! Using radar, scientists were able to "see" beneath the huge dunes of the desert, where they finally found the lost city of Ubar.

Science, Technology, and Society

Geocaching

Wouldn't it be exciting to go on a hunt for buried treasure? Thousands of people around the world participate in geocaching, which is an adventure game for GPS users. In this adventure game, individuals and groups of people put caches, or hidden treasures, in places all over the world. Once the cache is hidden, the coordinates of the cache's location are posted on the Internet. Then, geocaching teams compete to find the cache. Geocaching should only be attempted with parental supervision.

Roads appear as purple lines on this computer-generated remote-sensing image.

Language Arts ACTIVITY

Why was the word *geocaching* chosen for this adventure game? Use the Internet or another source to find the origin and meaning of the word *geocaching*.

Social Studies ACTIVITY

WRITING SKILL Ubar was once a very wealthy, magnificent city. Its riches were built on the frankincense trade. Research the history of frankincense, and write a paragraph describing how frankincense was used in ancient times and how it is used today.

Answer to Language Arts Activity

The word *geocaching* comes from the Greek root *geo*, meaning "Earth," and the word *cache*, which is a type of hiding place for money, food, or other necessities.

Answer to Social Studies Activity

Sample answer: Frankincense was used to treat illnesses and disguise body odor. Ancient civilizations from Rome to India treasured frankincense. Today, frankincense is used mainly as an incense or an herbal scent in some commercial products.

Matthew Henson

Arctic Explorer Matthew Henson was born in Maryland in 1866. His parents were freeborn sharecroppers. When Henson was a young boy, his parents died. He then went to look for work as a cabin boy on a ship. Several years later, Henson had traveled around the world and had become educated in the areas of geography, history, and mathematics. In 1898, Henson met U.S. Naval Lieutenant Robert E. Peary. Peary was the leader of Arctic expeditions between 1886 and 1909.

Peary asked Henson to accompany him as a navigator on several trips, including trips to Central America and Greenland. One of Peary's passions was to be the first person to reach the North Pole. It was Henson's vast knowledge of mathematics and carpentry that made Peary's trek to the North Pole possible. In 1909, Henson was the first person to reach the North Pole. Part of Henson's job as navigator was to drive ahead of the party and blaze the first trail. As a result, he often arrived ahead of everyone else. On April 6, 1909, Henson reached the approximate North Pole 45 minutes ahead of Peary. Upon his arrival, he exclaimed, "I think I'm the first man to sit on top of the world!"

Math ACTIVITY

On the last leg of their journey, Henson and Peary traveled 664.5 km in 16 days! On average, how far did Henson and Peary travel each day?

To learn more about these Science in Action topics, visit **go.hrw.com** and type in the keyword **HZ5MAPF**.

Current Science
Check out Current Science® articles related to this chapter by visiting **go.hrw.com.** Just type in the keyword **HZ5CS02.**

Answer to Math Activity
664.5 km ÷ 16 days = 41.53 km/day.

Apart from enduring subfreezing temperatures, sudden snow-storms, and slow starvation, Peary's team had to deal with the unique conditions of ice sheets that cover the Arctic Ocean. Movements of water currents under the ice cause constant changes on its surface. These changes include "pressure ridges," or small, steep mountains of ice that well up on the surface, and "leads," or open lanes of water caused from drifts or rents in the ice. Twice, Henson saved Peary's life by pulling him out of the freezing water of a suddenly formed lead.

ACTiViTY ———— GENERAL

The Explorers Club has been an international meeting place for explorers and scientists since 1904. Some of the most famous and influential field researchers in the world have been invited to join its ranks. Have students research past members of this organization. Then, have them plot on a map all the places these members have explored or discovered. (Students will find that some of the famous members of the Explorers Club include Tenzing Norgay and Sir Edmund Hillary, the first people to climb Mount Everest, Theodore Roosevelt, 26th President and founding member, Roald Amundsen, the first to reach the South Pole and sail the Northwest Passage, Herbert Hoover, 34th President, and Richard Byrd, the first aviator to fly over the Antarctic.)

Weathering and Soil Formation
Chapter Planning Guide

Compression guide:
To shorten instruction because of time limitations, omit the Chapter Lab.

OBJECTIVES	LABS, DEMONSTRATIONS, AND ACTIVITIES	TECHNOLOGY RESOURCES
PACING • 90 min pp. 30–37 **Chapter Opener**	SE **Start-up Activity,** p. 31 ◆ GENERAL	OSP **Parent Letter** ■ GENERAL CD **Student Edition on CD-ROM** CD **Guided Reading Audio CD** ■ TR **Chapter Starter Transparency*** VID **Brain Food Video Quiz**
Section 1 Weathering • Describe how ice, water, wind, gravity, plants, and animals cause mechanical weathering. • Describe how water, acids, and air cause chemical weathering of rocks.	TE **Group Activity** Identifying Weathering, p. 32 GENERAL TE **Group Activity** Acid Precipitation, p. 35 ADVANCED TE **Activity** CO_2 and Rain, p. 34 GENERAL SE **Quick Lab** Acids React!, p. 36 ◆ GENERAL CRF **Datasheet for Quick Lab*** SE **Model-Making Lab** Rockin' Through Time, p. 52 ◆ GENERAL CRF **Datasheet for Chapter Lab*** LB **Whiz-Bang Demonstrations** When it Rains, It Fizzes* ◆ GENERAL LB **EcoLabs & Field Activities** Whether It Weathers (or Not)* ◆ GENERAL	CRF **Lesson Plans*** TR **Bellringer Transparency*** TR **Chemical Weathering of Granite*** TR **LINK TO PHYSICAL SCIENCE** pH Values of Common Materials* VID **Lab Videos for Earth Science** CD **Science Tutor**
PACING • 45 min pp. 38–41 **Section 2 Rates of Weathering** • Explain how the composition of rock affects the rate of weathering. • Describe how a rock's total surface area affects the rate at which the rock weathers. • Describe how differences in elevation and climate affect the rate of weathering.	TE **Group Activity** Surface Area and Weathering, p. 38 ◆ GENERAL SE **School-to-Home Activity** Ice Wedging, p. 40 GENERAL LB **Calculator-Based Labs** How Low Can You Go?* ◆ ADVANCED TE **Activity** Differential Weathering, p. 39 ◆ ADVANCED	CRF **Lesson Plans*** TR **Bellringer Transparency*** TR **Total Surface Area to Volume*** CD **Science Tutor**
PACING • 45 min pp. 42–47 **Section 3 From Bedrock to Soil** • Describe the source of soil. • Explain how the different properties of soil affect plant growth. • Describe how various climates affect soil.	TE **Group Activity** Describing Soil, p. 42 ◆ GENERAL TE **Connection Activity** Environmental Science, p. 43 GENERAL TE **Group Activity** Living Soil, p. 44 ◆ ADVANCED LB **Calculator-Based Labs** A Hot and Cool Lab* ◆ ADVANCED LB **Calculator-Based Labs** A Soil Study* ◆ ADVANCED TE **Connection Activity** Biology, p. 44 GENERAL TE **Activity** Soil Layers, p. 45 ◆ GENERAL	CRF **Lesson Plans*** TR **Bellringer Transparency*** TR **Soil Horizons*** SE **Internet Activity,** p. 46 GENERAL CD **Science Tutor**
PACING • 45 min pp. 48–51 **Section 4 Soil Conservation** • Describe three important benefits that soil provides. • Describe four methods of preventing soil damage and loss.	TE **Activity** Uses of Soil, p. 49 BASIC TE **Connection Activity** Math, p. 49 GENERAL SE **Science in Action** Math, Social Studies, and Language Arts Activities, pp. 58–59 GENERAL LB **Long-Term Projects & Research Ideas** Precious Soil* ADVANCED	CRF **Lesson Plans*** TR **Bellringer Transparency*** CRF **SciLinks Activity*** GENERAL CD **Science Tutor**

PACING • 90 min

CHAPTER REVIEW, ASSESSMENT, AND STANDARDIZED TEST PREPARATION

CRF **Vocabulary Activity*** GENERAL
SE **Chapter Review,** pp. 54–55 GENERAL
CRF **Chapter Review*** ■ GENERAL
CRF **Chapter Tests A*** ■ GENERAL, **B*** ADVANCED, **C*** SPECIAL NEEDS
SE **Standardized Test Preparation,** pp. 56–57 GENERAL
CRF **Standardized Test Preparation*** GENERAL
CRF **Performance-Based Assessment*** GENERAL
OSP **Test Generator** GENERAL
CRF **Test Item Listing*** GENERAL

Online and Technology Resources

Visit **go.hrw.com** for a variety of free resources related to this textbook. Enter the keyword **HZ5WSF**.

Students can access interactive problem-solving help and active visual concept development with the *Holt Science and Technology* Online Edition available at **www.hrw.com**.

 Guided Reading Audio CD
Also in Spanish

A direct reading of each chapter for auditory learners, reluctant readers, and Spanish-speaking students.

 Science Tutor CD-ROM

Excellent for remediation and test practice.

SKILLS DEVELOPMENT RESOURCES	SECTION REVIEW AND ASSESSMENT	CORRELATIONS
SE Pre-Reading Activity, p. 30 GENERAL **OSP** Science Puzzlers, Twisters & Teasers GENERAL		National Science Education Standards UCP 2; SAI 1; SPSP 5; ES 2a
CRF Directed Reading A* ■ BASIC, B* SPECIAL NEEDS **CRF** Vocabulary and Section Summary* ■ GENERAL **SE** Reading Strategy Paired Summarizing, p. 32 GENERAL **SE** Connection to Chemistry Acidity of Precipitation, p. 35 GENERAL **TE** Inclusion Strategies, p. 35 ◆ **CRF** Reinforcement Worksheet Autobiography of a Rock* BASIC	**SE** Reading Checks, pp. 33, 34, 37 GENERAL **TE** Reteaching, p. 36 BASIC **TE** Quiz, p. 36 GENERAL **TE** Alternative Assessment, p. 36 GENERAL **SE** Section Review,* p. 37 ■ GENERAL **CRF** Section Quiz* ■ GENERAL	SAI 1, 2; ES 1c, 1d, 1k; *Chapter Lab:* UCP 2; *LabBook:* UCP 2; SAI 1
CRF Directed Reading A* ■ BASIC, B* SPECIAL NEEDS **CRF** Vocabulary and Section Summary* ■ GENERAL **SE** Reading Strategy Reading Organizer, p. 38 GENERAL	**SE** Reading Checks, pp. 39, 40, 41 GENERAL **TE** Reteaching, p. 40 BASIC **TE** Quiz, p. 40 GENERAL **TE** Alternative Assessment, p. 40 GENERAL **SE** Section Review,* p. 41 ■ GENERAL **CRF** Section Quiz* ■ GENERAL	SAI 1, 2; ES 1c, 1d
CRF Directed Reading A* ■ BASIC, B* SPECIAL NEEDS **CRF** Vocabulary and Section Summary* ■ GENERAL **SE** Reading Strategy Prediction Guide, p. 42 GENERAL **TE** Inclusion Strategies, p. 43 ◆ **SE** Connection to Social Studies Deforestation in Brazil, p. 45 GENERAL	**SE** Reading Checks, pp. 42, 45, 46 GENERAL **TE** Reteaching, p. 46 BASIC **TE** Quiz, p. 46 GENERAL **TE** Alternative Assessment, p. 46 GENERAL **SE** Section Review,* p. 47 ■ GENERAL **CRF** Section Quiz* ■ GENERAL	SAI 1, 2; SPSP 4; ES 1c, 1e, 1g, 1k
CRF Directed Reading A* ■ BASIC, B* SPECIAL NEEDS **CRF** Vocabulary and Section Summary* ■ GENERAL **SE** Reading Strategy Reading Organizer, p. 48 GENERAL **SE** Math Practice Making Soil, p. 49 GENERAL **CRF** Reinforcement Worksheet Where the Tall Corn Grows* BASIC **CRF** Critical Thinking Buying the Farm* ADVANCED	**SE** Reading Checks, pp. 48, 51 GENERAL **TE** Homework, p. 50 ADVANCED **TE** Reteaching, p. 50 BASIC **TE** Quiz, p. 50 GENERAL **TE** Alternative Assessment, p. 50 GENERAL **SE** Section Review,* p. 51 ■ GENERAL **CRF** Section Quiz* ■ GENERAL	SAI 1; SPSP 2, 4, 5; ST 2; HNS 1

One-Stop Planner® CD-ROM

This CD-ROM package includes:
- Lab Materials QuickList Software
- Holt Calendar Planner
- Customizable Lesson Plans
- Printable Worksheets
- ExamView® Test Generator
- Interactive Teacher Edition
- Holt PuzzlePro® Resources
- Holt PowerPoint® Resources

SCILINKS® NSTA

www.scilinks.org

Maintained by the **National Science Teachers Association.** See Chapter Enrichment pages for a complete list of topics.

Current Science®

Check out **Current Science** articles and activities by visiting the HRW Web site at **go.hrw.com.** Just type in the keyword **HZ5CS10T.**

Classroom Videos

- **Lab Videos** demonstrate the chapter lab.
- **Brain Food Video Quizzes** help students review the chapter material.
- **CNN Videos** bring science into your students' daily life.

Visual Resources

CHAPTER STARTER TRANSPARENCY

BELLRINGER TRANSPARENCIES

Section: Weathering
How do you think potholes form in paved roads? Write a few sentences that describe how water contributes to the formation of potholes. Illustrate how cycles of freezing and thawing cause potholes to grow.

Record your answers in your **science journal**.

Section: Rates of Weathering
Imagine that you are in a sand castle-building competition at the beach. Describe a variety of ways to protect your castle against the weathering effects of the wind and waves. Which of your inventions might be built on a large scale to help actual coastal communities? Are there any potential dangers to the environment that could be caused by such protective devices?

Write and illustrate your answers in your **science journal**.

TEACHING TRANSPARENCIES

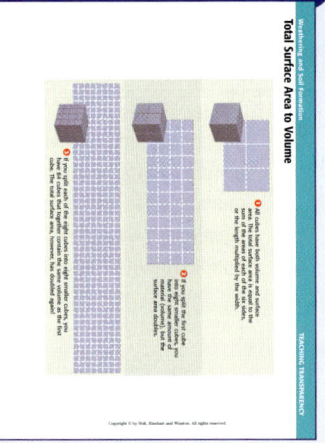

TEACHING TRANSPARENCIES

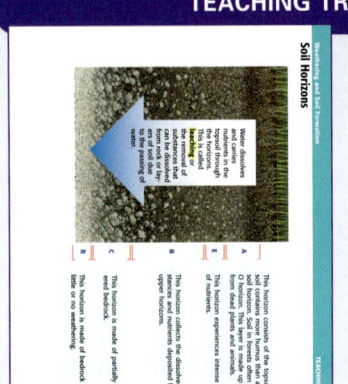

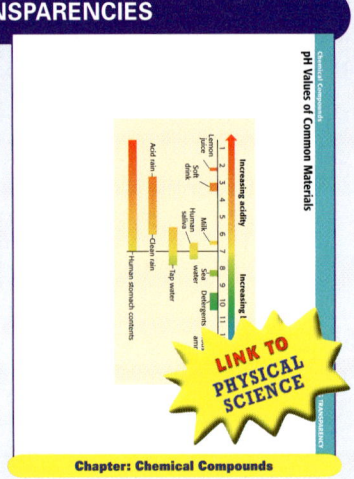

CONCEPT MAPPING TRANSPARENCY

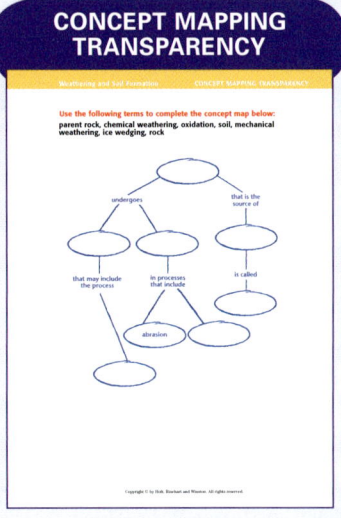

Planning Resources

LESSON PLANS

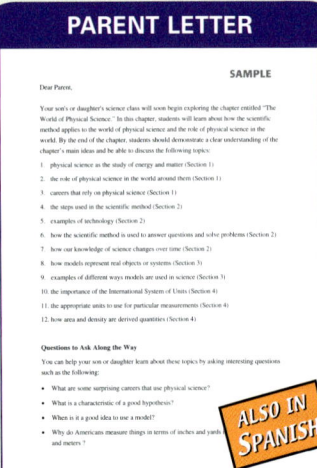

PARENT LETTER

ALSO IN SPANISH

TEST ITEM LISTING

One-Stop Planner® CD-ROM

This CD-ROM includes all of the resources shown here and the following time-saving tools:

- *Lab Materials QuickList Software*
- *Customizable lesson plans*
- *Holt Calendar Planner*
- *The powerful ExamView® Test Generator*

For a preview of available worksheets covering math and science skills, see pages T12–T19. All of these resources are also on the One-Stop Planner®.

Meeting Individual Needs

DIRECTED READING A
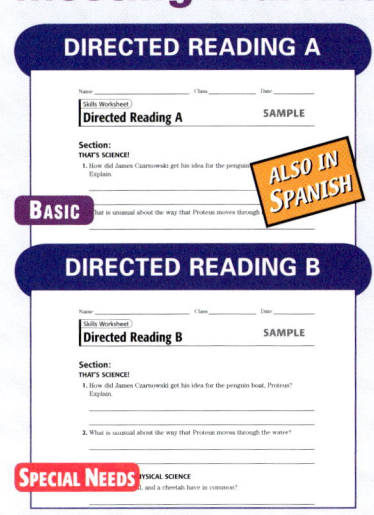
BASIC · ALSO IN SPANISH

VOCABULARY ACTIVITY
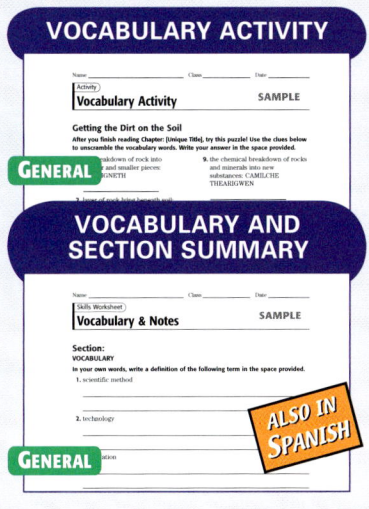
GENERAL

REINFORCEMENT
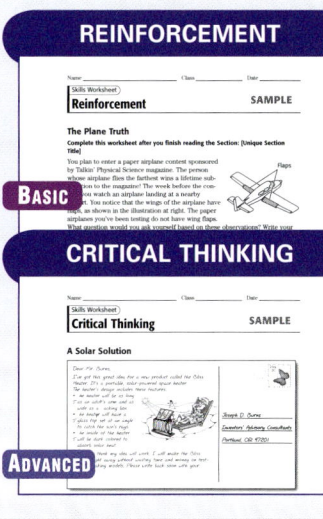
BASIC

SCILINKS ACTIVITY

GENERAL

DIRECTED READING B
SPECIAL NEEDS

VOCABULARY AND SECTION SUMMARY
GENERAL · ALSO IN SPANISH

CRITICAL THINKING
ADVANCED

SCIENCE PUZZLERS, TWISTERS & TEASERS
GENERAL

Labs and Activities

ECOLABS & FIELD ACTIVITIES

GENERAL

WHIZ-BANG DEMONSTRATIONS
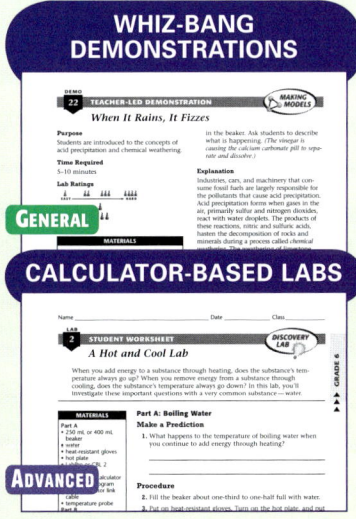
GENERAL

CALCULATOR-BASED LABS
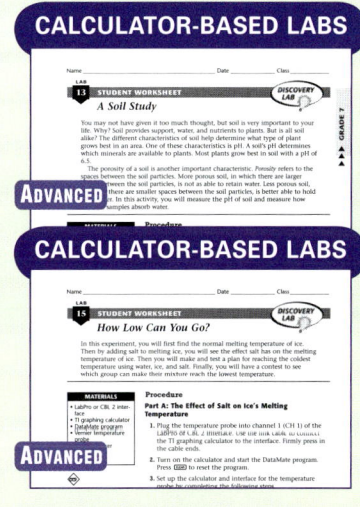
ADVANCED

DATASHEETS FOR QUICK LABS
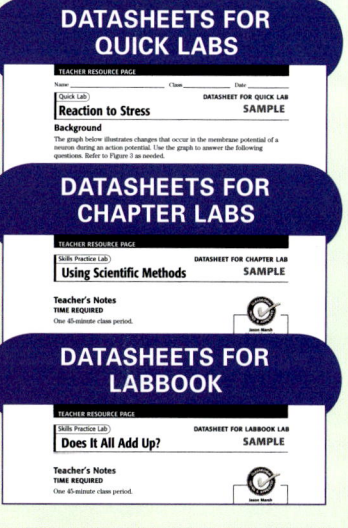

LONG-TERM PROJECTS & RESEARCH IDEAS
ADVANCED

CALCULATOR-BASED LABS
ADVANCED

CALCULATOR-BASED LABS
ADVANCED

DATASHEETS FOR CHAPTER LABS

DATASHEETS FOR LABBOOK

Review and Assessments

SECTION QUIZ
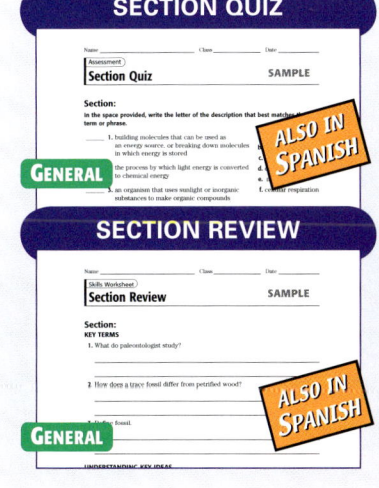
GENERAL · ALSO IN SPANISH

CHAPTER REVIEW
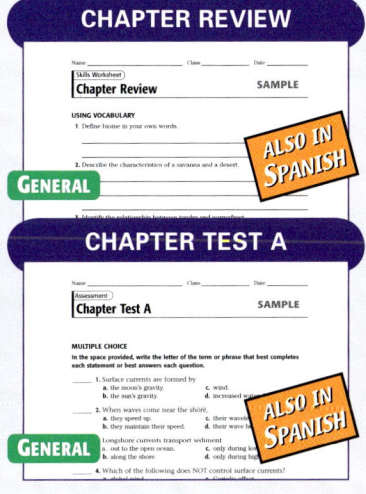
GENERAL · ALSO IN SPANISH

CHAPTER TEST B
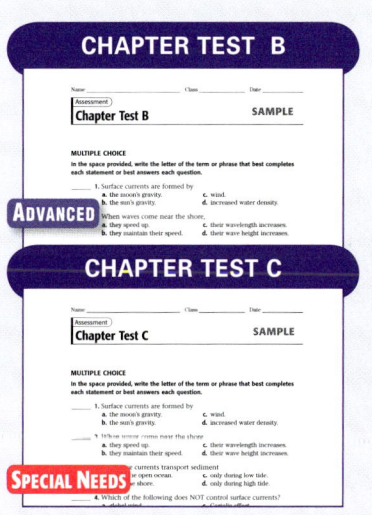
ADVANCED

STANDARDIZED TEST PREPARATION
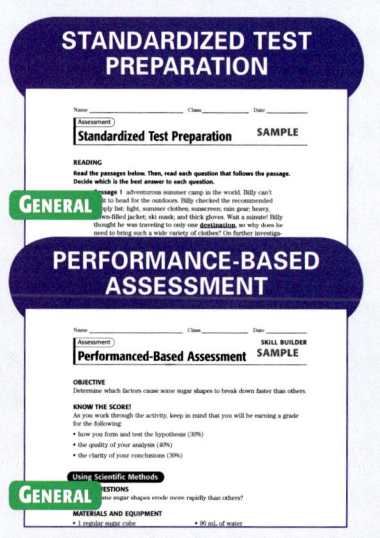
GENERAL

SECTION REVIEW
GENERAL · ALSO IN SPANISH

CHAPTER TEST A
GENERAL · ALSO IN SPANISH

CHAPTER TEST C
SPECIAL NEEDS

PERFORMANCE-BASED ASSESSMENT
GENERAL

This Chapter Enrichment provides relevant and interesting information to expand and enhance your presentation of the chapter material.

Section 1

Weathering

Thermal Contraction and Expansion

● There is much scientific debate over whether the daily and seasonal heating and cooling of rocks cause wide-scale weathering. In desert environments, where temperature ranges can be extreme, small rocks can shatter from expansion and contraction. But does this type of weathering occur in larger rocks and in climates that are more temperate? Geologists attempting to replicate this process in a lab have had little success. In one experiment, granite samples were repeatedly heated and cooled by more than 100°C, and no fracturing was observed. This suggests that if thermal expansion and contraction weathers rock, it may do so over the course of hundreds of thousands of years.

Salt Cracking

● In places where groundwater contains dissolved salts, salt water seeps into bedrock. When the water evaporates, the dissolved salts crystallize, and the growing crystals can exert enough force to fracture rock. This process, known as *salt cracking*, can be seen at the ocean, where sea cliffs become pitted and cracked from salt deposits. In desert regions, salt cracking erodes the base of some sandstone formations, which leaves an unweathered rock balancing on an eroded pedestal.

Weathered Mountain

● Mount Fuji is a dormant volcano that is a source of national pride among the Japanese. Unfortunately, the forces of mechanical weathering threaten to change the volcano's conical shape and near-perfect symmetry. To preserve the mountain's shape, the Japanese government built a 17 m–long concrete brace over a widening crevice near the mountain's summit. Before action was taken, as much as 300,000 tons of rock and soil had fallen down the mountainside every year.

Section 2

Rates of Weathering

Mineral Composition and Weathering Rates

● The order in which minerals crystallize from magma is nearly the same as the order in which they weather. Minerals that form quickly and at high temperatures and pressures within Earth, such as olivine and pyroxene, tend to be unstable at the surface and are less resistant to chemical weathering. Minerals that form slowly and at lower temperatures are much more resistant to the effects of weathering.

Section 3

From Bedrock to Soil

Types of Soil in the United States

- The soils of the mainland United States can be divided into two major types—pedocal and pedalfer. Pedocal is a calcium-rich soil that covers most of the western United States. Pedocal gets its name from the Latin *ped,* meaning "soil," joined with *cal,* representing *calcium.* Pedalfer is an iron- and aluminum-rich soil that covers most of the eastern half of the country. The *al* in *pedalfer* stands for *aluminum;* the *fer* stands for *ferrum* (iron).

Salinization

- All groundwater contains small concentrations of salts. If arid or semiarid soil is intensively irrigated, it can accumulate so much salt that it cannot support plant life. This process, called *salinization,* can ruin croplands. Some historical scholars argue that salinization contributed to the decline and fall of many ancient societies, including the Babylonian civilization.

Is That a Fact!

- ◆ *Regolith* is a term that describes all of the weathered material that lies over the bedrock. Soil refers to the upper layers of the regolith that supports plant life.

- ◆ The term regolith is derived from the Greek word rhegos, meaning "blanket," and the Greek word lithos meaning "stone." This derivation is important because it denotes the protective qualities of soil. Like a blanket, the soil protects the rock below from weathering. In mountain regions where soil is easily eroded, bedrock weathers much more quickly.

Section 4

Soil Conservation

Farming in the Imperial Valley

- Although desert soils are low in organic matter, they are not necessarily poor soils. Desert soil such as that of the Imperial Valley in California is actually quite rich with the minerals needed for plant growth. Water diverted from the Colorado River is used to irrigate the valley, which is now one of the nation's major farming regions, where crops such as alfalfa, cotton, and sugar beets are grown. While the Imperial Valley is incredibly productive, agriculture in the region relies heavily on the use of fertilizers. There is also much debate over whether the Imperial Valley diverts too much water from the Colorado River.

Federal Soil Conservation Service

- In response to the devastating windstorms that swept across the Great Plains, the U.S. Department of Agriculture formed the Soil Conservation Service in 1935. Working with ranchers and farmers, conservationists instituted strategies such as contour plowing and terracing, planting trees as windbreaks, allowing land to lie fallow, and planting drought-resistant crops.

SCILINKS

NSTA
Developed and maintained by the
National Science Teachers Association

SciLinks is maintained by the National Science Teachers Association to provide you and your students with interesting, up-to-date links that will enrich your classroom presentation of the chapter.

Visit www.scilinks.org and enter the SciLinks code for more information about the topic listed.

Topic: Weathering
SciLinks code: HSM1648

Topic: Rates of Weathering
SciLinks code: HSM1269

Topic: Soil and Climate
SciLinks code: HSM1408

Topic: Soil Types
SciLinks code: HSM1412

Topic: Soil Conservation
SciLinks code: HSM1409

Overview

Tell students that this chapter will help them learn about the process of weathering, including factors that cause weathering and factors that effect the rate of weathering. Students will learn about how soil is formed and how the properties of soil affect plant growth. They will also learn about the effect of climate on soil. Finally students will learn about soil conservation.

Assessing Prior Knowledge

Students should be familiar with the following topics:

- acids in precipitation, in groundwater, and in vegetation
- various types of rocks
- chemical reactions

Identifying Misconceptions

As students learn the material in this chapter, some of them may associate soil formation with deposits by rivers. Other students may think that the soil has existed since the Earth formed. Discuss the components of soil, and discuss where those components came from.

Weathering and Soil Formation

About the PHOTO

Need a nose job, Mr. President? The carving of Thomas Jefferson that is part of the Mount Rushmore National Memorial is having its nose inspected by a National Parks worker. The process of weathering has caused cracks to form in the carving of President Jefferson. National Parks workers use a sealant to protect the memorial from moisture, which can cause further cracking.

PRE-READING ACTIVITY

FOLDNOTES **Key-Term Fold** Before you read the chapter, create the FoldNote entitled "Key-Term Fold" described in the **Study Skills** section of the Appendix. Write a key term from the chapter on each tab of the key-term fold. Under each tab, write the definition of the key term.

Standards Correlations

National Science Education Standards

The following codes indicate the National Science Education Standards that correlate to this chapter. The full text of the standards is at the front of the book.

Chapter Opener
UCP 2; SAI 1; SPSP 5; ES 2a

Section 1 Weathering
SAI 1; ES 1c, 1k; *LabBook*: UCP 2; SAI 1

Section 2 Rates of Weathering
SAI 1; ES 1c

Section 3 From Bedrock to Soil
ES 1e, 1k

Section 4 Soil Conservation
SPSP 2, 4, 5

Chapter Lab
UCP 2; SAI 1

Chapter Review
UCP 1; SPSP 2; HNS 1; ES 1c, 1e, 1k

Science in Action
ST 2; HNS 3; ES 1k

START-UP ACTIVITY
MATERIALS

FOR EACH GROUP
- container, small (2)
- spoon (2)
- stopwatch or timepiece with second hand
- sugar, granulated (1 tsp)
- sugar cube
- water

Teacher's Notes: The size of sugar cubes may vary, so students may obtain more accurate results if they use two sugar cubes rather than 1 sugar cube and 1 tsp of granulated sugar. Students can crush one of the sugar cubes between two spoons and then compare the rate at which both sugar samples dissolve.

Answers

1. Sample answer: The granulated sugar dissolved faster than the sugar cube because the grains of sugar had more surface area than the sugar cube did. Therefore, the water could come into contact with and dissolve the granulated sugar more quickly.

2. Sample answer: Several smaller rocks would wear away faster because they have more surface area than a large rock does.

START-UP ACTIVITY

What's the Difference?

In this chapter, you will learn about the processes and rates of weathering. Complete this activity to learn about how the size and surface area of a substance affects how quickly the substance breaks down.

Procedure

1. Fill **two small containers** about half full with **water.**

2. Add **one sugar cube** to one container.

3. Add **1 tsp of granulated sugar** to the other container.

4. Using **one spoon for each container,** stir the water and sugar in each container at the same rate.

5. Using a **stopwatch,** measure how long it takes for the sugar to dissolve in each container.

Analysis

1. Did the sugar dissolve at the same rate in both containers? Explain why or why not.

2. Do you think one large rock or several smaller rocks would wear away faster? Explain your answer.

Chapter Starter Transparency
Use this transparency to help students begin thinking about the processes that weather rock.

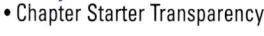

Focus

Overview

In this section, students will learn how processes such as ice wedging and abrasion and plant and animal activities contribute to the mechanical weathering of rock. Students will also learn how water and acids cause chemical weathering of rock.

Bellringer

Ask students to think about how potholes form in paved roads. Have students write a few sentences that describe how water contributes to the formation of potholes. Students should illustrate how cycles of freezing and thawing help cause potholes to grow.

Motivate

Group ACTIVITY — GENERAL

Identifying Weathering Have groups of students find photographs in magazines that illustrate weathering. Examples may include rusted cars or bikes, sidewalks or walls that have been cracked by plant roots, potholes, and weathered statues. Ask the class to help you group the photographs into examples of mechanical weathering and chemical weathering. **LS Logical**

READING WARM-UP

Objectives

- Describe how ice, water, wind, gravity, plants, and animals cause mechanical weathering.
- Describe how water, acids, and air cause chemical weathering of rocks.

Terms to Learn

weathering
mechanical weathering
abrasion
chemical weathering
acid precipitation

READING STRATEGY

Paired Summarizing Read this section silently. In pairs, take turns summarizing the material. Stop to discuss ideas that seem confusing.

weathering the process by which rock materials are broken down by the action of physical and chemical processes

mechanical weathering the breakdown of rock into smaller pieces by physical means

Weathering

If you have ever walked along a trail, you might have noticed small rocks lying around. Where did these rocks come from?

These smaller rocks came from larger rocks that were broken down. **Weathering** is the process by which rock materials are broken down by the action of physical or chemical processes.

Mechanical Weathering

If you were to crush one rock with another rock, you would be demonstrating one type of mechanical weathering. **Mechanical weathering** is the breakdown of rock into smaller pieces by physical means. Agents of mechanical weathering include ice, wind, water, gravity, plants, and even animals.

Ice

The alternate freezing and thawing of soil and rock, called *frost action,* is a form of mechanical weathering. One type of frost action, *ice wedging,* is shown in **Figure 1.** Ice wedging starts when water seeps into cracks during warm weather. When temperatures drop, the water freezes and expands. The ice then pushes against the sides of the crack. This causes the crack to widen.

Figure 1   Ice Wedging

The granite in the photo has been broken down by repeated ice wedging, which is shown below.

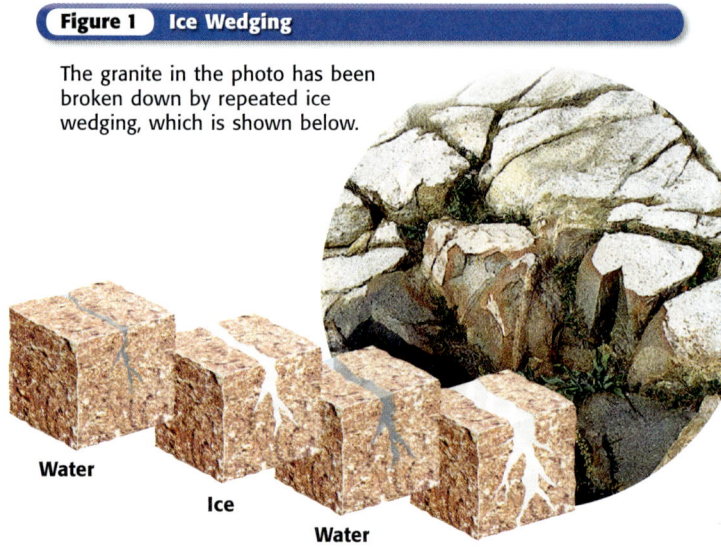

Water

Ice

Water

Ice

CHAPTER RESOURCES

Chapter Resource File

- Lesson Plan
- Directed Reading A **BASIC**
- Directed Reading B **SPECIAL NEEDS**

Technology

Transparencies
- Bellringer

Figure 2 Three Forms of Abrasion

These river rocks are rounded because they have been tumbled in the riverbed by fast-moving water for many years.

This rock has been shaped by blowing sand. Such rocks are called ventifacts.

Rocks grind against each other in a rock slide, which creates smaller and smaller rock fragments.

Abrasion

As you scrape a piece of chalk against a board, particles of the chalk rub off to make a line on the board and the piece of chalk wears down and becomes smaller. The same process, called *abrasion,* happens with rocks. **Abrasion** is the grinding and wearing away of rock surfaces through the mechanical action of other rock or sand particles.

Wind, Water, and Gravity

Abrasion can happen in many ways, as shown in **Figure 2.** When rocks and pebbles roll along the bottom of swiftly flowing rivers, they bump into and scrape against each other. The weathering that occurs eventually causes these rocks to become rounded and smooth.

Wind also causes abrasion. When wind blows sand and silt against exposed rock, the sand eventually wears away the rock's surface. The figure above (center) shows what this kind of sandblasting can do to a rock.

Abrasion also occurs when rocks fall on one another. You can imagine the forces rocks exert on each other as they tumble down a mountainside. In fact, anytime one rock hits another, abrasion takes place.

✓ **Reading Check** Name three things that can cause abrasion.
(See the Appendix for answers to Reading Checks.)

abrasion the grinding and wearing away of rock surfaces through the mechanical action of other rock or sand particles

CONNECTION to
Physical Science — ADVANCED

Exfoliation Another process of mechanical weathering is called *exfoliation.* As overlying rock is removed by uplift and erosion, the pressure on the rock is reduced. As the pressure is reduced, the rock expands in volume and long, curved cracks develop parallel to the rock's surface. In this way, an outcrop "sheds" layers of rock. Exfoliation can often be observed in granite outcrops. Ask students to describe why granite formations are prone to exfoliation. (Granite forms underground, so it forms under a great deal of pressure from the rock above. A granite pluton 15 km underground forms at 5,000 times the pressure at Earth's surface. As the granite is pushed toward the surface and the overlying rock is weathered away, the pressure on the rock is reduced and the granite exfoliates.) **LS** **Verbal**

Answer to Reading Check
Wind, water, and gravity can cause abrasion.

Cultural Awareness — GENERAL

The Ajanta Caves In the second century BCE, Buddhist monks began carving an intricate system of caves in a massive basalt flow in central India. The Ajanta caves comprised a complex of monasteries, temples, and living quarters. The caves were adorned with beautiful frescoes and carvings and then were mysteriously abandoned in the seventh century CE. They were rediscovered by British game hunters less than 200 years ago. The Ajanta caves are notable not only for their artwork but also for the manner in which they were carved. The monks cut channels in the rock first and then jammed dry logs into the crevices. They poured water on top of the logs and waited for the expanding wood to shatter the rock. In this way, they carved 30 caves out of solid rock.

Humans Cause Weathering
Students may be surprised to learn that animals such as earthworms, coyotes, and rabbits play significant roles in weathering rock. Human activity also contributes to the weathering of rock. People move large amounts of soil and rock whenever they farm, build, or drive off-road vehicles. In addition, people blast rock to make tunnels, roads, mines, and quarries.

CONNECTION to Life Science — GENERAL

Gold Bugs As ground-dwelling termites construct their homes, they excavate an enormous amount of soil and rock fragments. Occasionally, the termites strike it rich. Geochemical prospectors have learned from indigenous cultures in Africa, Asia, Australia, and South America to analyze termite mounds for ore deposits such as tin, silver, gold, diamond, and uranium. In some parts of Africa, gold concentrations in termite mounds are rich enough that people earn money by panning gold from the mounds.

Answer to Reading Check
Answers may vary. Sample answer: ants, worms, mice, coyotes, and rabbits.

Plants

You may not think of plants as being strong, but some plants can easily break rocks. Have you ever seen sidewalks and streets that are cracked because of tree roots? Roots don't grow fast, but they certainly are powerful! Plants often send their roots into existing cracks in rocks. As the plant grows, the force of the expanding root becomes so strong that the crack widens. Eventually, the entire rock can split apart, as shown in **Figure 3**.

Animals

Believe it or not, earthworms cause a lot of weathering! They burrow through the soil and move soil particles around. This exposes fresh surfaces to continued weathering. Would you believe that some kinds of tropical worms move an estimated 100 metric tons of soil per acre every year? Almost any animal that burrows causes mechanical weathering. Ants, worms, mice, coyotes, and rabbits are just some of the animals that contribute to weathering. **Figure 4** shows some of these animals in action. The mixing and digging that animals do often contribute to another type of weathering, called *chemical weathering*. You will learn about this type of weathering next.

✓ **Reading Check** List three animals that can cause weathering.

Figure 3 *Although they grow slowly, tree roots are strong enough to break solid rock.*

Figure 4 *Animals that live in the soil, such as moles, prairie dogs, insects, worms, and gophers, cause a lot of weathering. When the animals burrow in the ground, they break up soil and loosen rocks to be exposed to further weathering.*

ACTIVITY — GENERAL

CO$_2$ and Rain Have students try this activity to learn how CO$_2$ combines with water in the atmosphere to form a slightly acidic solution. Fill a test tube halfway with water. Add a few drops of universal indicator solution. Have a student exhale through a straw into the water. As the CO$_2$ combines with the water, carbonic acid forms and the color of the solution changes. This color change indicates an acidic solution. **LS Visual/Kinesthetic**

Figure 5 Chemical Weathering of Granite

After thousands of years of chemical weathering, even hard rock, such as granite, can turn to sediment.

1 Rain, weak acids, and air chemically weather granite.

2 The bonds between mineral grains weaken as weathering proceeds.

3 When granite is weathered, it makes sand and clay, also called sediment.

Chemical Weathering

The process by which rocks break down as a result of chemical reactions is called **chemical weathering.** Common agents of chemical weathering are water, weak acids, and air.

Water

If you drop a sugar cube into a glass of water, the sugar cube will dissolve after a few minutes. This process is an example of chemical weathering. Even hard rock, such as granite, can be broken down by water. But, it just may take thousands of years. **Figure 5** shows how granite is chemically weathered.

Acid Precipitation

Rain, sleet, or snow, that contains a high concentration of acids is called **acid precipitation.** Precipitation is naturally acidic. However, acid precipitation contains more acid than normal precipitation. The high level of acidity can cause very rapid weathering of rock. Small amounts of sulfuric and nitric acids from natural sources, such as volcanoes, can make precipitation acidic. However, acid precipitation can also be caused by air pollution from the burning of fossil fuels, such as coal and oil. When these fuels are burned, they give off gases, including sulfur oxides, nitrogen oxides, and carbon oxides. When these compounds combine with water in the atmosphere, they form weak acids, which then fall back to the ground in rain and snow. When the acidity is too high, acid precipitation can be harmful to plants and animals.

chemical weathering the process by which rocks break down as a result of chemical reactions

acid precipitation rain, sleet, or snow, that contains a high concentration of acids

CONNECTION TO Chemistry

Acidity of Precipitation
Acidity is measured by using a pH scale, the units of which range from 0 to 14. Solutions that have a pH of less than 7 are acidic. Research some recorded pH levels of acid rain. Then, compare these pH levels with the pH levels of other common acids, such as lemon juice and acetic acid.

INCLUSION Strategies

• **Attention Deficit Disorder**   • **Learning Disabled**
• **Developmentally Delayed**

Organize students into small groups. To each group, hand out a cork, sandpaper, and a small plastic container with a lid such as a margarine container. Ask students to abrade their "rock" by sanding away some of the cork. Next, students should completely fill the plastic containers with water. Place the containers in a freezer until the next class. Students should predict what will happen to the water-filled containers in their **science journal.** The next class, ask students to relate this experiment to the real world and discuss how they modeled two forms of mechanical weathering.  **Kinesthetic**

Figure 6 Acid in groundwater has weathered limestone to form Carlsbad Caverns, in New Mexico.

Mechanical or Chemical?
To reinforce the difference between chemical and mechanical weathering, have students decide whether each of the following phenomena is an example of mechanical or chemical weathering:

- a rock fall on a mountainside (mechanical)
- a rusty bridge (chemical)
- lichens and mosses growing on a boulder (chemical)
- an alpine glacier advancing down a valley (mechanical)

English Language Learners

LS Verbal

Quiz — GENERAL

1. How do earthworms aid in weathering? (When earthworms burrow, they move soil particles around and expose fresh surfaces to weathering.)

2. What human activities can increase the acidity of precipitation? (activities that burn fossil fuels, such as coal)

Alternative Assessment — GENERAL

Trivia Challenge Divide students into small groups. Have each group research the process of weathering. Ask each group to create five multiple-choice trivia cards for a game that tests the players' knowledge of weathering. **LS Kinesthetic**

Acids in Groundwater

In certain places groundwater contains weak acids, such as carbonic or sulfuric acid. These acids react with rocks in the ground, such as limestone. When groundwater comes in contact with limestone, a chemical reaction occurs. Over a long period of time, the dissolving of limestone forms karst features, such as caverns. The caverns, like the one shown in **Figure 6,** form from the eating away of the limestone.

Acids in Living Things

Another source of acids that cause weathering might surprise you. Take a look at the lichens in **Figure 7.** Lichens produce acids that can slowly break down rock. If you have ever taken a walk in a park or forest, you have probably seen lichens growing on the sides of trees or rocks. Lichens can also grow in places where some of the hardiest plants cannot. For example, lichens can grow in deserts, in arctic areas, and in areas high above timberline, where even trees don't grow.

Figure 7 Lichens, which consist of fungi and algae living together, contribute to chemical weathering.

Acids React!

1. Ketchup is one example of a food that contains weak acids, which react with certain substances. Take a **penny** that has a dull appearance, rub **ketchup** on it for several minutes.

2. Rinse the penny.
3. Where did all the grime on the penny go?
4. How is this process similar to what happens to a rock when it is exposed to natural acids during weathering?

MATERIALS

FOR EACH GROUP
- ketchup
- penny

Safety Caution: Students who are allergic to tomatoes should use a cotton swab to apply the ketchup.

Answers

3. Answers may vary. Students might note that the grime on the surface of the penny reacted chemically with the acid in the ketchup and dissolved.

4. Answers may vary. Students should note that the way that rocks react with acids is similar to the way that the grime on the surface of the penny reacted with the ketchup.

Air

The car shown in **Figure 8** is undergoing chemical weathering due to the air. The oxygen in the air is reacting with the iron in the car, causing the car to rust. Water speeds up the process. But the iron would rust even if no water were present. Scientists call this process oxidation.

Oxidation is a chemical reaction in which an element, such as iron, combines with oxygen to form an oxide. This common form of chemical weathering is what causes rust. Old cars, aluminum cans, and your bike can experience oxidation if left exposed to air and rain for long periods of time.

✔ **Reading Check** What can cause oxidation?

Figure 8 *Rust is a result of chemical weathering.*

SECTION Review

Summary

- Ice wedging is a form of mechanical weathering in which water seeps into rock cracks and then freezes and expands.
- Wind, water, and gravity cause mechanical weathering by abrasion.
- Animals and plants cause mechanical weathering by turning the soil and breaking apart rocks.
- Water, acids, and air chemically weather rock by weakening the bonds between mineral grains of the rock.

Using Key Terms

1. In your own words, write a definition for each of the following terms: *weathering, mechanical weathering, abrasion, chemical weathering* and *acid precipitation.*

Understanding Key Ideas

2. Which of the following things cannot cause mechanical weathering?
   a. water
   b. acid
   c. wind
   d. animals

3. List three things that cause chemical weathering of rocks.

4. Describe three ways abrasion occurs in nature.

5. Describe the similarity in the ways tree roots and ice mechanically weather rock.

6. Describe five sources of chemical weathering.

Critical Thinking

7. **Making Inferences** Why does acid precipitation weather rocks faster than normal precipitation?

8. **Making Comparisons** Compare the weather processes that affect a rock on top of a mountain and a rock buried beneath the ground.

Math Skills

9. Substances that have a pH of less than 7 are acidic. For each pH unit lower, the acidity is ten times greater. For example, normal precipitation is slightly acidic at a 5.6 pH. If acid precipitation were measured at 4.6 pH, it would be 10 times more acidic than normal precipitation. How many times more acidic would precipitation at 3.6 pH be than normal precipitation?

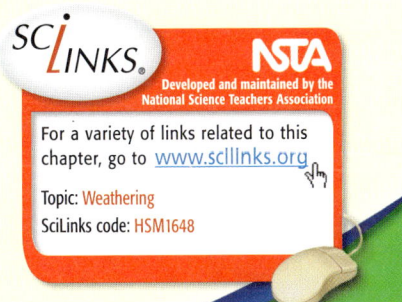

SCI LINKS
NSTA
Developed and maintained by the National Science Teachers Association

For a variety of links related to this chapter, go to www.scilinks.org

Topic: Weathering
SciLinks code: HSM1648

CHAPTER RESOURCES

Chapter Resource File

- Section Quiz GENERAL
- Section Review GENERAL
- Vocabulary and Section Summary GENERAL
- Reinforcement Worksheet BASIC
- Datasheet for Quick Lab

Focus

Overview

This section explores how different types of rock, climate, and elevation affect weathering rates.

🔔 Bellringer

Ask students to imagine that they are in a sand castle–building competition at the beach. Ask them to come up with ways to protect their castle against the weathering effects of the wind and waves. Students can share their ideas with the class.

Motivate

Group ACTIVITY — GENERAL

Surface Area and Weathering

Supply each group with a clear glass that contains a calcium antacid tablet and a second glass that contains a calcium antacid tablet cut into quarters. Tell students that both antacid tablets and limestone contain calcium carbonate, which dissolves in acidic solutions. Have students pour enough vinegar into the glasses to cover the tablets. Ask students which of the tablets "weathers" more rapidly. (The tablet cut into quarters weathers more rapidly.) Lead students to conclude that surface area affects the rate at which materials weather. **LS Logical**

READING WARM-UP

Objectives

● Explain how the composition of rock affects the rate of weathering.

● Describe how a rock's total surface area affects the rate at which the rock weathers.

● Describe how differences in elevation and climate affect the rate of weathering.

Terms to Learn

differential weathering

READING STRATEGY

Reading Organizer As you read this section, create an outline of the section. Use the headings from the section in your outline.

Rates of Weathering

Have you ever seen a cartoon in which a character falls off a cliff and lands on a ledge? Ledges exist in nature because the rock that the ledge is made of weathers more slowly than the surrounding rock.

Weathering is a process that takes a long time. However, some rock will weather faster than other rock. The rate at which a rock weathers depends on climate, elevation, and the makeup of the rock.

Differential Weathering

Hard rocks, such as granite, weather more slowly than softer rocks, such as limestone. **Differential weathering** is a process by which softer, less weather resistant rocks wear away and leave harder, more weather resistant rocks behind.

Figure 1 shows a landform that has been shaped by differential weathering. Devils Tower was once a mass of molten rock deep inside an active volcano. When the molten rock cooled and hardened, it was protected from weathering by the outer rock of the volcano. After thousands of years of weathering, the soft outer parts of the volcano have worn away. The harder, more resistant rock is all that remains.

Figure 1 *The illustration is an artist's idea of how the original volcano may have looked. The photo inset shows Devils Tower as it appears today.*

CHAPTER RESOURCES

Chapter Resource File

• Lesson Plan
• Directed Reading A BASIC
• Directed Reading B SPECIAL NEEDS

Technology

Transparencies
• Bellringer
• Total Surface Area to Volume

The Shape of Rocks

Weathering takes place on the outer surface of rocks. Therefore, the more surface area that is exposed to weathering, the faster the rock will be worn down. A large rock has a large surface area. But a large rock also has a large volume. Because of the large rock's volume, the large rock will take a long time to wear down.

If a large rock is broken into smaller fragments, weathering of the rock happens much more quickly. The rate of weathering increases because a smaller rock has more surface area to volume than a larger rock has. So, more of a smaller rock is exposed to the weathering process. **Figure 2** shows this concept in detail.

differential weathering the process by which softer, less weather resistant rocks wear away and leave harder, more weather resistant rocks behind

✔️ **Reading Check** How does an increase in surface area affect the rate of weathering? (*See the Appendix for answers to Reading Checks.*)

Figure 2 Total Surface Area to Volume

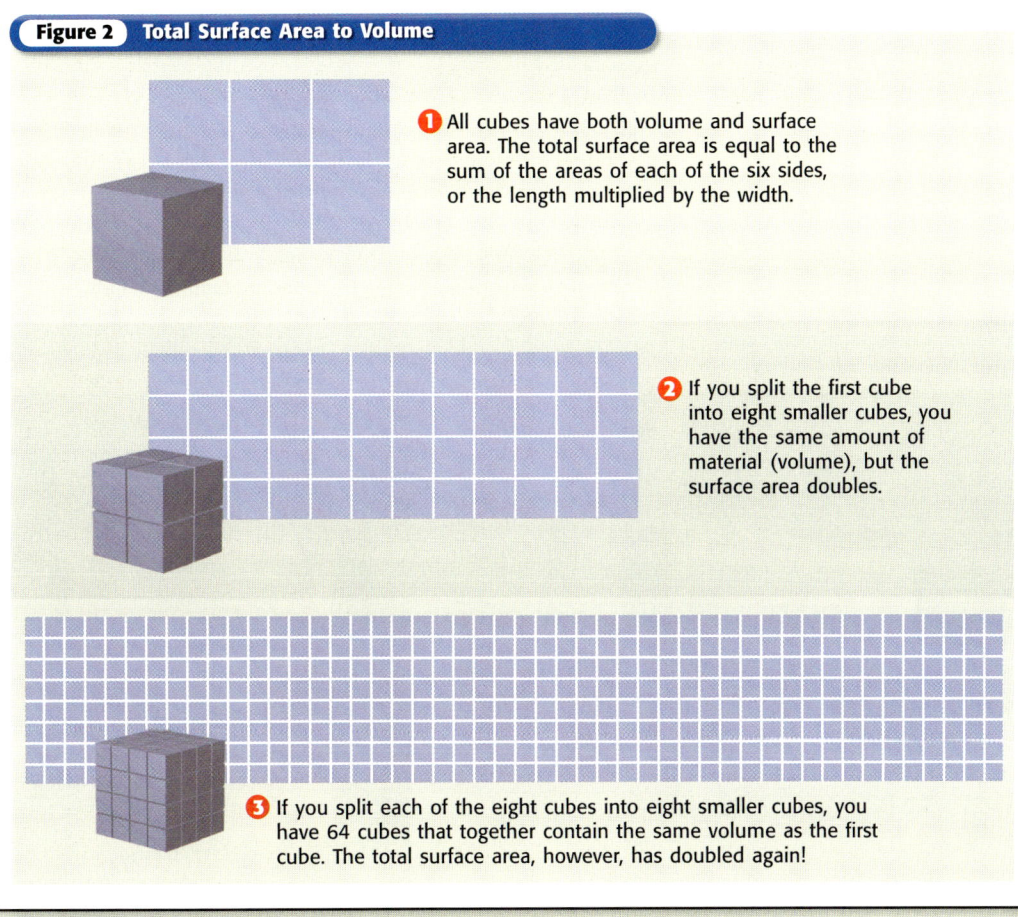

❶ All cubes have both volume and surface area. The total surface area is equal to the sum of the areas of each of the six sides, or the length multiplied by the width.

❷ If you split the first cube into eight smaller cubes, you have the same amount of material (volume), but the surface area doubles.

❸ If you split each of the eight cubes into eight smaller cubes, you have 64 cubes that together contain the same volume as the first cube. The total surface area, however, has doubled again!

Teach

ACTIVITY ━━━━ **ADVANCED**

Differential Weathering Have students use the Internet to learn more about how differential weathering created landforms in the United States. Students could investigate certain locations where differential weathering has created spectacular landforms, such as Good City of Rocks in Gooding, Idaho, and The Window in Big Bend National Park, Texas. **LS Visual**

MISCONCEPTION ALERT

Weathering of Hard Rocks Students may assume that some types of rock, such as granite, do not weather. Emphasize that all rocks weather, but different kinds of rock weather at different rates. The granite that is used in buildings and monuments is often polished. Polishing the surface slows the weathering process because less surface area is exposed.

Answer to Reading Check
As the surface area increases, the rate of weathering also increases.

CONNECTION to Geology ━━━━ GENERAL

Weathering Devils Tower Why is the intrusive volcanic rock that makes up Devils Tower more resistant to weathering than was the extrusive volcanic rock of the former volcano? Both rock types had the same composition. The difference is their cohesiveness. The intrusive rock cooled more slowly than the extrusive rock. As the rock cooled, it formed large crystals that interlocked like a 3-D jigsaw puzzle. As a result, the volcanic neck was more resistant to weathering. In contrast, the rock that made up the outside of the volcano cooled quickly. This rock was made of much smaller crystals and groundmass material—material that cooled so fast that it did not form crystals.

Weathering Rates Show students a variety of rocks. Ask students what they could do to increase the rate of weathering of the rocks. (Sample answer: The rocks could be on top of a high mountain; the rocks could be put in a warm, humid climate; and the rocks could be crushed to increase the surface area.) **LS Visual**

Quiz — GENERAL

1. Do different types of rock weather at different rates? (yes)

2. Does chemical weathering affect the rate of mechanical weathering. (yes)

3. What factors contribute to accelerated weathering rates at high elevations? (wind, precipitation, and gravity)

Alternative Assessment — GENERAL

Investigate Your Area A cemetery is a great place to observe the effects of differential weathering because several kinds of rock are used to make headstones and most of the headstones are dated. Schedule a field trip to a cemetery, or encourage interested students to visit a cemetery on their own. Have them compare the dates and types of rock used to determine which kinds of rock are most susceptible to weathering. **LS Kinesthetic**

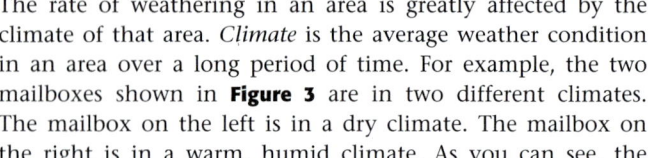

Ice Wedging

WRITING SKILL To understand ice wedging, try this activity at home with a parent. Fill a small, plastic water bottle with water. Plug the opening with a piece of putty. Place the bottle in the freezer overnight. Describe in your **science journal** what happened to the putty.

Figure 3 These photos show the effects different climates can have on rates of weathering.

◀ This mailbox is in a dry climate and does not experience a high rate of weathering.

This mailbox is in a warm, humid climate. It experiences a high rate of chemical weathering called oxidation. ▶

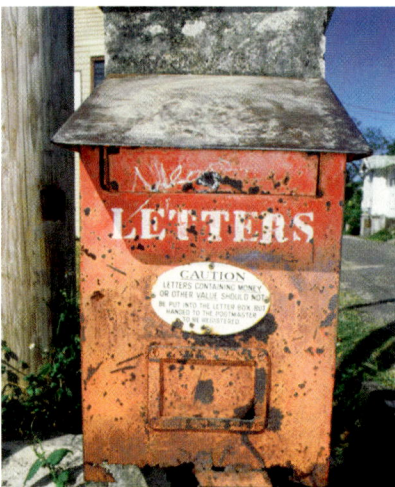

Weathering and Climate

The rate of weathering in an area is greatly affected by the climate of that area. *Climate* is the average weather condition in an area over a long period of time. For example, the two mailboxes shown in **Figure 3** are in two different climates. The mailbox on the left is in a dry climate. The mailbox on the right is in a warm, humid climate. As you can see, the mailbox in the warm, humid climate is rusty.

Temperature and Water

The rate of chemical weathering happens faster in warm, humid climates. The rusty mailbox has experienced a type of chemical weathering called oxidation. Oxidation, like other chemical reactions, happens at a faster rate when temperatures are higher and when water is present.

Water also increases the rate of mechanical weathering. The freezing of water that seeps into the cracks of rocks is the process of ice wedging. Ice wedging causes rocks to break apart. Over time, this form of weathering can break down even the hardest rocks into soil.

Temperature is another major factor in mechanical weathering. The more often temperatures cause freezing and thawing, the more often ice wedging takes place. Therefore, climatic regions that experience frequent freezes and thaws have a greater rate of mechanical weathering.

✓ **Reading Check** Why would a mailbox in a warm, humid climate experience a higher rate of weathering than a mailbox in a cold, dry climate?

Answer to School-to-Home Activity

Sample answer: The putty was pushed out of the opening of the bottle when the water in the bottle froze and expanded.

Answer to Reading Check

Warm, humid climates have higher rates of weathering because oxidation happens faster when temperatures are higher and when water is present.

Weathering and Elevation

Just like everything else, mountains are exposed to air and water. As a result, mountain ranges are weathered down. Weathering happens on mountains in the same way it does everywhere else. However, as shown in **Figure 4,** rocks at higher elevations, as on a mountain, are exposed to more wind, rain, and ice than the rocks at lower elevations are. This increase in wind, rain, and ice at higher elevations causes the peaks of mountains to weather faster.

Gravity affects weathering, too. The steepness of mountain slopes increases the effects of mechanical and chemical weathering. Steep slopes cause rainwater to quickly run off the sides of mountains. The rainwater carries the sediment down the mountain's slope. This continual removal of sediment exposes fresh rock surfaces to the effects of weathering. New rock surfaces are also exposed to weathering when gravity causes rocks to fall away from the sides of mountains. The increased surface area means weathering happens at a faster rate.

✓ Reading Check Why do mountaintops weather faster than rocks at sea level?

Figure 4 The ice, rain, and wind that these mountain peaks are exposed to cause them to weather at a fast rate.

SECTION Review

Summary

- Hard rocks weather more slowly than softer rocks.
- The more surface area of a rock that is exposed to weathering, the faster the rock will be worn down.
- Chemical weathering occurs faster in warm, humid climates.
- Weathering occurs faster at high elevations because of an increase in ice, rain, and wind.

Using Key Terms

1. In your own words, write a definition for the term *differential weathering*.

Understanding Key Ideas

2. A rock will have a lower rate of weathering when the rock
   a. is in a humid climate.
   b. is a very hard rock, such as granite.
   c. is at a high elevation.
   d. has more surface area exposed to weathering.

3. How does surface area affect the rate of weathering?

4. How does climate affect the rate of weathering?

5. Why does the peak of a mountain weather faster than the rocks at the bottom of the mountain?

Math Skills

6. The surface area of an entire cube is 96 cm². If the length and width of each side are equal, what is the length of one side of the cube?

Critical Thinking

7. **Making Inferences** Does the rate of chemical weathering increase or stay the same when a rock becomes more mechanically weathered? Why?

For a variety of links related to this chapter, go to www.scilinks.org

Topic: Rates of Weathering
SciLinks code: HSM1269

Answer to Reading Check

Mountains weather faster because they are exposed to more wind, rain, and ice, which are agents of weathering.

CHAPTER RESOURCES

Chapter Resource File

- Section Quiz **GENERAL**
- Section Review **GENERAL**
- Vocabulary and Section Summary **GENERAL**

Overview

In this section, students will learn about sources for soil formation, the various properties of soil, and the effects of climate on soil type.

🔔 Bellringer

Have students answer the following questions:

- Has there always been soil on Earth? (No, soil did not exist until the parent rock of the early Earth was weathered.)

- What makes soil valuable to humans? (Answers may vary. Soil supports the growth of plants, which provide humans with oxygen and food.)

Motivate

Group ACTIVITY — GENERAL

Describing Soil Provide small groups of students with magnifying lenses and samples of several types of local soil. Have students empty each sample onto a piece of white paper and examine it. Have students record their observations about each sample's composition, color, particle size, texture, and moisture content. Ask groups to hypothesize how each soil type formed and what type of plant life might grow in the soil. **LS Kinesthetic** Co-op Learning

---

SECTION
3

READING WARM-UP

Objectives

- Describe the source of soil.
- Explain how the different properties of soil affect plant growth.
- Describe how various climates affect soil.

Terms to Learn

soil   soil structure
parent rock   humus
bedrock   leaching
soil texture

READING STRATEGY

Prediction Guide Before you read this section, write the title of each heading in this section. Next, under each heading, write what you think you will learn.

soil a loose mixture of rock fragments, organic material, water, and air that can support the growth of vegetation

parent rock a rock formation that is the source of soil

bedrock the layer of rock beneath soil

From Bedrock to Soil

Most plants need soil to grow. But what exactly is soil? Where does it come from?

The Source of Soil

To a scientist, **soil** is a loose mixture of small mineral fragments, organic material, water, and air that can support the growth of vegetation. But not all soils are the same. Because soils are made from weathered rock fragments, the type of soil that forms depends on the type of rock that weathers. The rock formation that is the source of mineral fragments in the soil is called **parent rock.**

Bedrock is the layer of rock beneath soil. In this case, the bedrock is the parent rock because the soil above it formed from the bedrock below. Soil that remains above its parent rock is called *residual soil.*

Soil can be blown or washed away from its parent rock. This soil is called *transported soil.* **Figure 1** shows one way that soil is moved from one place to another. Both wind and the movement of glaciers are also responsible for transporting soil.

✓ **Reading Check** What is soil formed from? (*See the Appendix for answers to Reading Checks.*)

Figure 1 *Transported soil may be moved long distances from its parent rock by rivers, such as this one.*

CHAPTER RESOURCES

Chapter Resource File

- **Lesson Plan**
- **Directed Reading A** BASIC
- **Directed Reading B** SPECIAL NEEDS

Technology

📦 **Transparencies**
- Bellringer

Answer to Reading Check

Soil is formed from parent rock, organic material, water, and air.

Figure 2 **Soil Texture**

The proportion of these different-sized particles in soil determine the soil's texture.

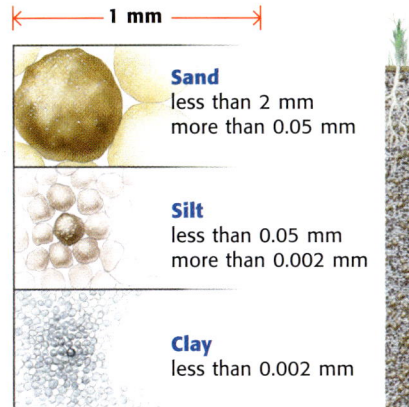

← 1 mm →

Sand
less than 2 mm
more than 0.05 mm

Silt
less than 0.05 mm
more than 0.002 mm

Clay
less than 0.002 mm

This callout shows the makeup of sandy loam. It is made of
Sand 60%
Silt 30%
Clay 10%

Soil Properties

Some soils are great for growing plants. Other soils can't support the growth of plants. To better understand soil, you will next learn about its properties, such as soil texture, soil structure, and soil fertility.

Soil Texture and Soil Structure

Soil is made of different-sized particles. These particles can be as large as 2 mm, such as sand. Other particles can be too small to see without a microscope. **Soil texture** is the soil quality that is based on the proportions of soil particles. **Figure 2** shows the soil texture for a one type of soil.

Soil texture affects the soil's consistency. Consistency describes a soil's ability to be worked and broken up for farming. For example, soil texture that has a large proportion of clay can be hard and difficult for farmers to break up.

Soil texture influences the *infiltration,* or ability of water to move through soil. Soil should allow water to get to the plants' roots without causing the soil to be completely saturated.

Water and air movement through soil is also influenced by soil structure. **Soil structure** is the arrangement of soil particles. Soil particles are not always evenly spread out. Often, one type of soil particle will clump in an area. A clump of one type of soil can either block water flow or help water flow, which affects soil moisture.

soil texture the soil quality that is based on the proportions of soil particles.

soil structure the arrangement of soil particles

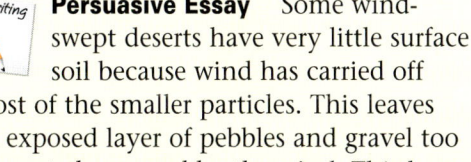

MATERIALS

FOR EACH GROUP
- bag, plastic, resealable
- bread, without preservatives (several slices)
- soil samples (including potting soil)
- spatula and tongs

Living Soil Bacteria and fungi are the major decomposers of organic material in soil. These decomposers break down organic matter into simpler substances that plants can absorb. Have groups bring in several types of soil. Have students put 20 to 30 drops of distilled water on each of the bread slices. Using the spatula, students should then sprinkle onto each slice of bread a small amount of each soil sample. Next, students should use the tongs to place each slice of bread in a separate, labeled plastic bag. One slice of bread on which there is no soil should be used as a control. Place the bags in a dark box or drawer. After five to seven days, students may observe, describe, and analyze the patterns of mold grown in each sample by using a hand lens. Remind students that the bags must remain sealed during the observation period. Finally, ask the students to draw conclusions based on the data that they have recorded for each sample. **LS Kinesthetic**

humus the dark, organic material formed in soil from the decayed remains of plants and animals

leaching the removal of substances that can be dissolved from rock, ore, or layers of soil due to the passing of water

Soil Fertility

Nutrients in soil, such as iron, are necessary for plants to grow. Some soils are rich in nutrients. Other soils may not have many nutrients or are not able to supply the nutrients to the plants. A soil's ability to hold nutrients and to supply nutrients to a plant is described as *soil fertility*. Many nutrients in soil come from the parent rock. Other nutrients come from **humus,** which is the organic material formed in soil from the decayed remains of plants and animals. These remains are broken down into nutrients by decomposers, such as bacteria and fungi.

Soil Horizons

Because of the way soil forms, soil often ends up in a series of layers, with humus-rich soil on top, sediment below that, and bedrock on the bottom. Geologists call these layers *horizons*. The word *horizon* tells you that the layers are horizontal. **Figure 3** shows what these horizons can look like. You can see these layers in some road cuts.

The top layer of soil is often called the *topsoil*. Topsoil contains more humus than the layers below it. The humus is rich in the nutrients plants need to be healthy. This is why good topsoil is necessary for farming.

Figure 3 Soil Horizons

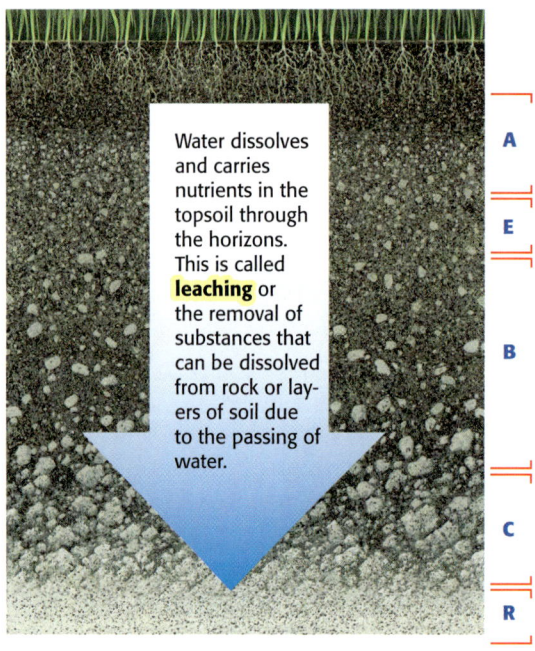

Water dissolves and carries nutrients in the topsoil through the horizons. This is called **leaching** or the removal of substances that can be dissolved from rock or layers of soil due to the passing of water.

A This horizon consists of the topsoil. Topsoil contains more humus than any other soil horizon. Soil in forests often has an O horizon. The O horizon is made up of litter from dead plants and animals.

E This horizon experiences intense leaching of nutrients.

B This horizon collects the dissolved substances and nutrients deposited from the upper horizons.

C This horizon is made of partially weathered bedrock.

R This horizon is made of bedrock that has little or no weathering.

Is That a Fact!

Where do soda cans and airplanes come from? They come from the soil, of course! In tropical areas, the process of soil leaching produces concentrated bauxite deposits in a thin layer at the Earth's surface. Bauxite is the ore that is refined to produce aluminum.

CONNECTION ACTIVITY
Biology — GENERAL

Berlese Funnels Students may be surprised to learn that soil is a thriving ecosystem. In 1 m³ of soil, there may be 10 million roundworms and 50,000 small insects and mites. In a single gram of fertile soil, there may be 50,000 algae, 400,000 fungi, and 2.5 million bacteria. Have students construct a Berlese funnel to collect small organisms from soil. Designs for Berlese funnels are available on the Internet. **LS Kinesthetic**

Soil pH

Soils can be acidic or basic. The pH scale is used to measure how acidic or basic a soil is and ranges from 0 to 14. The midpoint, which is 7, is neutral. Soil that has a pH below 7 is acidic. Soil that has a pH above 7 is basic.

The pH of a soil influences how nutrients dissolve in the soil. For example, plants are unable to take up certain nutrients from soils that are basic, or that have a high pH. Soils that have a low pH can restrict other important nutrients from hungry plants. Because different plants need different nutrients, the right pH for a soil depends on the plants growing in it.

Soil and Climate

Soil types vary from place to place. One reason for this is the differences in climate. As you read on, you will see that climate can make a difference in the types of soils that develop around the world.

Tropical Rain Forest Climates

Take a look at **Figure 4.** In tropical rain forest climates, the air is very humid and the land receives a large amount of rain. Because of warm temperatures, crops can be grown year-round. The warm soil temperature also allows dead plants and animals to decay easily. This provides rich humus to the soil.

Because of the lush plant growth, you may think that tropical rain forest soils are the most nutrient-rich in the world. However, tropical rain forest soils are nutrient poor. The heavy rains in this climate leach precious nutrients from the topsoil into deeper layers of soil. The result is that tropical topsoil is very thin. Another reason tropical rain forest soil is nutrient poor is that the lush vegetation has a great demand for nutrients. The nutrients that aren't leached away are quickly taken up by plants and trees that live off the soil.

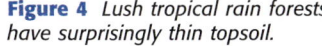 **Reading Check** Why is the topsoil in tropical rain forests thin?

Figure 4 *Lush tropical rain forests have surprisingly thin topsoil.*

CONNECTION TO Social Studies

WRITING SKILL **Deforestation in Brazil** In Brazil, rain forests have been cut down at an alarmingly high rate, mostly by farmers. However, tropical rain forest topsoil is very thin and is not suitable for long-term farming. Research the long-term effects of deforestation on the farmers and indigenous people of Brazil. Then, write a one page report on your findings.

Cultural Awareness **GENERAL**

PORTFOLIO **Sustainable Farming** The Lacandon Maya of Mexico have developed sustainable farming methods that do not destroy the fragile soil of the tropical rain forest. On a small piece of land, they grow both food crops and tree crops, a practice known as *agroforestry*. After a few years, they let the farmland recover by allowing it to become a forest again. The Lacandon Maya's approach to farming is recognized for its ecological soundness and has been replicated in many countries. Have students research sustainable farming techniques and create a model or poster to share with the class. **LS Visual**

Answer to Reading Check

Heavy rains leach precious nutrients into deeper layers of soil, resulting in a very thin layer of topsoil.

ACTIVITY ———— GENERAL

Soil Layers Have student groups collect two soil samples from the same area, one from the surface and one from 16 to 20 cm below the surface. Groups should fill two test tubes about one-quarter full with each soil sample and should add water until the tubes are three-quarters full. Have students gently shake the covered tubes for several minutes. Place the test tubes in a rack, and leave them overnight. Students should be able to observe the different compositions of layers that formed in the two test tubes. Soil components will settle according to weight. Have students measure the depth of each layer in the test tube and observe the color and size of the grains. Then, have students draw and label the soil layers. The surface soil will probably contain noticeably more humus than the below-surface soil will. **LS Kinesthetic**

Factors That Affect Soil Ask students to think of factors that influence the characteristics of soil. How does climate affect soil? Write factors on the board. (Factors might include parent rock, sources of organic material, and rainfall.) Be sure that students understand how these characteristics affect the soil's ability to sustain vegetation. **LS** Verbal

Quiz — GENERAL

1. What is the source of mineral fragments in soil? (parent rock)

2. What is the organic part of soil called? (humus)

3. What causes topsoil in tropical climates to be thin? (Sample answer: leaching from heavy rains)

Alternative Assessment — GENERAL

Writing **Making Postcards** Have students imagine that they are on a world trip during which they travel to every climate mentioned in the section. Tell them to write a series of postcards in which they describe what the soil is like in each climate. The picture on each card should be a magazine photograph that illustrates the soil in that climate. **LS** Visual

Answer to Reading Check

Temperate climates have the most productive soil.

Figure 5 *The salty conditions of desert soils make it difficult for many plants to survive.*

For another activity related to this chapter, go to **go.hrw.com** and type in the keyword **HZ5WSFW.**

Desert Climates

While tropical climates get a lot of rain, deserts get less than 25 cm a year. Leaching of nutrients is not a problem in desert soils. But the lack of rain causes many other problems, such as very low rates of chemical weathering and less ability to support plant and animal life. A low rate of weathering means soil is created at a slower rate.

Some water is available from groundwater. Groundwater can trickle in from surrounding areas and seep to the surface. But as soon as the water is close to the surface, it evaporates. So, any materials that were dissolved in the water are left behind in the soil. Without the water to dissolve the minerals, the plants are unable to take them up. Often, the chemicals left behind are various types of salts. These salts can sometimes become so concentrated that the soil becomes toxic, or poisonous, even to desert plants! Death Valley, shown in **Figure 5,** is a desert that has toxic levels of salt in the soil.

Temperate Forest and Grassland Climates

Much of the continental United States has a temperate climate. An abundance of weathering occurs in temperate climates. Temperate areas get enough rain to cause a high level of chemical weathering, but not so much that the nutrients are leached out of the soil. Frequent changes in temperature lead to frost action. As a result, thick, fertile soils develop, as shown in **Figure 6.**

Temperate soils are some of the most-productive soils in the world. In fact, the midwestern part of the United States has earned the nickname "breadbasket" for the many crops the region's soil supports.

✓ **Reading Check** Which climate has the most-productive soil?

Figure 6 *The rich soils in areas that have a temperate climate support a vast farming industry.*

Cryptogamic Soil In some desert areas, a special type of soil called *cryptogamic soil* is actually alive! This soil is composed of different species of mosses, lichens, fungi, and algae. Cryptogamic soil is sometimes known as "brown sugar soil" because it is dark brown and crusty. The spongy soil absorbs moisture readily and, when disturbed by freezing, it uplifts and cracks. The cracks are important to desert ecosystems because plant seeds get lodged in the cracks. The moisture allows the seeds to germinate. Cryptogamic soils can be severely damaged if they are walked on. Ask students to find out why walking on cryptogamic soil could damage the soil, and have them write a paragraph about it.

Arctic Climates

Arctic areas have so little precipitation that they are like cold deserts. In arctic climates, as in desert climates, chemical weathering occurs very slowly. So, soil formation also occurs slowly. Slow soil formation is why soil in arctic areas, as shown in **Figure 7,** is thin and unable to support many plants.

Arctic climates also have low soil temperatures. At low temperatures, decomposition of plants and animals happens more slowly or stops completely. Slow decomposition limits the amount of humus in the soil, which limits the nutrients available. These nutrients are necessary for plant growth.

Figure 7 *Arctic soils, such as the soil along Denali Highway, in Alaska, cannot support lush vegetation.*

SECTION Review

Summary

- Soil is formed from the weathering of bedrock.
- Soil texture affects how soil can be worked for farming and how well water passes through it.
- The ability of soil to provide nutrients so that plants can survive and grow is called *soil fertility.*
- The pH of a soil influences which nutrients plants can take up from the soil.
- Different climates have different types of soil, depending on the temperature and rainfall.

Using Key Terms

1. Use each of the following terms in a separate sentence: *soil, parent rock, bedrock, soil texture, soil structure, humus,* and *leaching.*

Understanding Key Ideas

2. Which of the following soil properties influences soil moisture?
   a. soil horizon
   b. soil fertility
   c. soil structure
   d. soil pH

3. Which of the following soil properties influences how nutrients can be dissolved in soil?
   a. soil texture
   b. soil fertility
   c. soil structure
   d. soil pH

4. When is parent rock the same as bedrock?

5. What is the difference between residual and transported soils?

6. Which climate has the most thick, fertile soil?

7. How does soil temperature influence arctic soil?

Math Skills

8. If a soil sample is 60% sand particles and has 30 million particles of soil, how many of those soil particles are sand?

Critical Thinking

9. **Identifying Relationships** In which type of climate would leaching be more common—tropical rain forest or desert?

10. **Making Comparisons** Although arctic climates are extremely different from desert climates, their soils may be somewhat similar. Explain why.

Developed and maintained by the National Science Teachers Association

For a variety of links related to this chapter, go to www.scilinks.org

Topic: Soil and Climate
SciLinks code: HSM1408

Answers to Section Review

1. Sample answer: Soil is made up of minerals, water, organic material, and air. Soil is formed from parent rock. Bedrock is the layer of rock under soil. Soil texture affects a soil's ability to be worked. Soil structure affects water and air movement through the soil. Humus helps make soil rich in nutrients. Leaching from heavy rains can remove important nutrients from the soil.

2. c

3. d

4. when the rock formation below the soil is also the source of mineral fragments in the soil

5. Residual soil remains above its parent rock. Transported soil is blown or washed away from its parent rock.

6. temperate forest and grassland climates

7. Low temperatures slow down decomposition, which limits the amount of humus in the soil.

8. 30 million $\times$ 60% = 18 million

9. tropical rain forests

10. Both are dry climates. Chemical weathering occurs more slowly in dry climates; therefore, both deserts and arctic climates experience less chemical weathering.

CHAPTER RESOURCES

Chapter Resource File

- Section Quiz GENERAL
- Section Review GENERAL
- Vocabulary and Section Summary GENERAL

Overview

In this section, students will learn the importance of soil. Students will then learn about the methods used to prevent nutrient loss and erosion of soil.

🔔 Bellringer

Tell students Franklin D. Roosevelt's quote: "The nation that destroys its soil destroys itself." Lead a discussion on the meaning of this quote.

Motivate

Discussion ——— GENERAL

Soil Engineering Students may think that all soils are merely dirt. Soils have different characteristics, which depend on soil composition. Engineers study soil types when planning roads and buildings. Different types of soils require different engineering considerations. For example, soils high in clay swell when they are wet and contract when they dry. The expanding and contracting can cause shifting and cracking in roadbeds and building foundations. Arrange for an engineer, contractor, or geologist to speak with the class about the importance of understanding soil types. Have students prepare questions to ask the guest speaker. **LS Auditory**

SECTION

4

Soil Conservation

Believe it or not, soil can be endangered, just like plants and animals. Because soil takes thousands of years to form, it is not easy to replace.

If we do not take care of our soils, we can ruin them or even lose them. Soil is a resource that must be conserved. **Soil conservation** is a method to maintain the fertility of the soil by protecting the soil from erosion and nutrient loss.

The Importance of Soil

Soil provides minerals and other nutrients for plants. If the soil loses these nutrients, then plants will not be able to grow. Take a look at the plants shown in **Figure 1.** The plants on the right look unhealthy because they are not getting enough nutrients. There is enough soil to support the plant's roots, but the soil is not providing them with the food they need. The plants on the left are healthy because the soil they live in is rich in nutrients.

All animals get their energy from plants. The animals get their energy either by eating the plants or by eating animals that have eaten plants. So, if plants can't get their nutrients from the soil, animals can't get their nutrients from plants.

✓ **Reading Check** Why is soil important? (*See the Appendix for answers to Reading Checks.*)

Housing

Soil also provides a place for animals to live. The region where a plant or animal lives is called its *habitat*. Earthworms, grubs, spiders, ants, moles, and prairie dogs all live in soil. If the soil disappears, so does the habitat for these animals.

Objectives

● Describe three important benefits that soil provides.

● Describe four methods of preventing soil damage and loss.

Terms to Learn

soil conservation
erosion

READING STRATEGY

Reading Organizer As you read this section, make a table comparing the four methods of preventing soil damage and loss.

soil conservation a method to maintain the fertility of the soil by protecting the soil from erosion and nutrient loss

Figure 1 *Both of these photos show the same crop, but the soil in the photo on the right is poor in nutrients.*

Answer to Reading Check

Soil provides nutrients to plants, houses for animals, and stores water.

Water Storage

Soil is also extremely important to plants for water storage. Without soil to hold water, plants would not get the moisture or the nutrients they need. Soil also keeps water from running off, flowing elsewhere, and possibly causing flooding.

Soil Damage and Loss

What would happen if there were no soil? Soil loss is a serious problem around the world. Soil damage can lead to soil loss. Soil can be damaged from overuse by poor farming techniques or by overgrazing. Overused soil can lose its nutrients and become infertile. Plants can't grow in soil that is infertile. Without plants to hold and help cycle water, the area can become a desert. This process, formally known as *desertification,* is called *land degradation*. Without plants and moisture, the soil can be blown or washed away.

Soil Erosion

When soil is left unprotected, it can be exposed to erosion. **Erosion** is the process by which wind, water, or gravity transport soil and sediment from one location to another. **Figure 2** shows Providence Canyon, which was formed from the erosion of soil when trees were cut down to clear land for farming. Roots from plants and trees are like anchors to the soil. Roots keep topsoil from being eroded. Therefore, plants and trees protect the soil. By taking care of the vegetation, you also take care of the soil.

MATH PRACTICE

Making Soil

Suppose it takes 500 years for 2 cm of new soil to form in a certain area. But the soil is eroding at a rate of 1 mm per year. Is the soil eroding faster than it can be replaced? Explain.

erosion the process by which wind, water, ice, or gravity transport soil and sediment from one location to another

Figure 2 Providence Canyon has suffered soil erosion from the cutting of forests for farmland.

Reteaching — BASIC

Soil Conservation On the board, write the headings "Soil erosion" on one side and "Nutrient depletion" on the other. Ask students to suggest different soil conservation methods, and have them state which problem (soil erosion or nutrient depletion) the method of conservation addresses. After a brief discussion of each method, write the method under the appropriate head. **LS** Verbal

Quiz — GENERAL

1. What is one way that nutrients are removed from soil?
   (Sample answer: by planting the same crops every year)

2. How do contour plowing and terracing help prevent soil erosion? (by interrupting water flow across the topsoil)

Alternative Assessment — GENERAL

Raising Awareness Declare "Soil Conservation Awareness Week." Have students create posters that alert your school to the importance of soil and that highlight some ways to protect and conserve soil. **LS** Visual

Figure 3 Soil Conservation Techniques

Contour plowing helps prevent erosion from heavy rains.

Terracing prevents erosion from heavy rains on steep hills.

No-till farming prevents erosion by providing cover that reduces water runoff.

Soybeans are a **cover crop** which restores nutrients to soil.

Contour Plowing and Terracing

If farmers plowed rows so that they ran up and down hills, what might happen during a heavy rain? The rows would act as river valleys and channel the rainwater down the hill, which would erode the soil. To prevent erosion in this way, a farmer could plow across the slope of the hills. This is called contour plowing. In *contour plowing,* the rows act as a series of dams instead of a series of rivers. **Figure 3** shows contour plowing and three other methods of soil conservation. If the hills are really steep, farmers can use *terracing.* Terracing changes one steep field into a series of smaller, flatter fields. *No-till farming,* which is the practice of leaving old stalks, provides cover from rain. The cover reduces water runoff and slows soil erosion.

Homework — ADVANCED

Indigenous Agriculture The ancient Maya of Central America used specialized agricultural techniques to maximize their corn crops. The Maya intentionally planted their crops over sinkholes. Soil over sinkholes is ideal because it is rich and all surface water drains into the sinkhole. Have students find out more about the agricultural innovations of other indigenous cultures and prepare a five-minute speech on the topic. Ask students if any of these techniques are still being used today. **LS** Verbal

Cover Crop and Crop Rotation

In the southern United States, during the early 1900s, the soil had become nutrient poor by the farming of only one crop, cotton. George Washington Carver, the scientist shown in **Figure 4,** urged farmers to plant soybeans and peanuts instead of cotton. Some plants, such as soybeans and peanuts, helped to restore important nutrients to the soil. These plants are called cover crops. *Cover crops* are crops that are planted between harvests to replace certain nutrients and prevent erosion. Cover crops prevent erosion by providing cover from wind and rain.

Another way to slow down nutrient depletion is through *crop rotation*. If the same crop is grown year after year in the same field, certain nutrients become depleted. To slow this process, a farmer can plant different crops. A different crop will use up less nutrients or different nutrients from the soil.

✓ Reading Check What can soybeans and peanuts do for nutrient-poor soil?

Figure 4 *George Washington Carver taught soil conservation techniques to farmers.*

SECTION Review

Summary

- Soil is important for plants to grow, for animals to live in, and for water to be stored.
- Soil erosion and soil damage can be prevented by contour plowing, terracing, using cover crop, and practicing crop rotation.

Using Key Terms

1. In your own words, write a definition for each of the following terms: *soil conservation* and *erosion*.

Understanding Key Ideas

2. What are three important benefits that soil provides?

3. Practicing which of the following soil conservation techniques will replace nutrients in the soil?
   a. cover crop use
   b. no-till farming
   c. terracing
   d. contour plowing

4. How does crop rotation benefit soil?

5. List four methods of soil conservation, and describe how each helps prevent the loss of soil.

Math Skills

6. Suppose it takes 500 years to form 2 cm of new soil without erosion. If a farmer needs at least 35 cm of soil to plant a particular crop, how many years will the farmer need to wait before planting his or her crop?

Critical Thinking

7. **Applying Concepts** Why do land animals, even meat eaters, depend on soil to survive?

SciLINKS

Developed and maintained by the National Science Teachers Association

For a variety of links related to this chapter, go to www.scilinks.org

Topic: Soil Conservation
SciLinks code: HSM1409

CONNECTION to History — GENERAL

George Washington Carver By the early 20th century, Southern cotton cultivation had so depleted soil nutrients that the area faced an agricultural crisis. George Washington Carver convinced farmers to plant peanuts and soybeans instead of cotton. These crops helped restore nitrogen to the soil. The soil recovered, and Carver's work helped revitalize the agricultural economy of the South. Ask students to learn more about the life of this remarkable scientist.

CHAPTER RESOURCES

Chapter Resource File
- Section Quiz GENERAL
- Section Review GENERAL
- Vocabulary and Section Summary GENERAL
- Reinforcement Worksheet BASIC
- Critical Thinking ADVANCED
- SciLinks Activity GENERAL

Rockin' Through Time

Teacher's Notes

Time Required

One 45-minute class period

Lab Ratings

EASY ———————→ HARD

Teacher Prep
Student Set-Up
Concept Level
Clean Up

MATERIALS

The materials listed on the student page are adequate for groups of 4–5 students.

Safety Caution

Remind students to review all safety cautions and icons before beginning this lab activity. Be sure to use plastic bottles in this activity.

OBJECTIVES

Design a model to understand how abrasion breaks down rocks.

Evaluate the effects of abrasion.

MATERIALS

- bottle, plastic, wide-mouthed, with lid, 3 L
- graph paper or computer
- markers
- pieces of limestone, all about the same size (24)
- poster board
- tap water

SAFETY

Rockin' Through Time

Wind, water, and gravity constantly change rocks. As wind and water rush over the rocks, the rocks may be worn smooth. As rocks bump against one another, their shapes change. The form of mechanical weathering that occurs as rocks collide and scrape together is called *abrasion*. In this activity, you will shake some pieces of limestone to model the effects of abrasion.

Ask a Question

1. How does abrasion break down rocks? How can I use this information to identify rocks that have been abraded in nature?

Form a Hypothesis

2. Formulate a hypothesis that answers the questions above.

Test the Hypothesis

3. Copy the chart on the next page onto a piece of poster board. Allow enough space to place rocks in each square.

4. Lay three of the limestone pieces on the poster board in the area marked "0 shakes." Be careful not to bump the poster board after you have added the rocks.

5. Place the remaining 21 rocks in the 3 L bottle. Then, fill the bottle halfway with water.

6. Close the lid of the bottle securely. Shake the bottle vigorously 100 times.

7. Remove three rocks from the bottle, and place them on the poster board in the box that indicates the number of times the rocks have been shaken.

8. Repeat steps 6 and 7 six times until all of the rocks have been added to the board.

CHAPTER RESOURCES

Chapter Resource File

- Datasheet for Chapter Lab
- Lab Notes and Answers

Technology

 Classroom Videos
- Lab Video

- Great Ice Escape

Analyze the Results

1 **Examining Data** Describe the surface of the rocks that you placed in the area marked "0 shakes." Are they smooth or rough?

2 **Describing Events** How did the shape of the rocks change as you performed this activity?

3 **Constructing Graphs** Using graph paper or a computer, construct a graph, table, or chart that describes how the shapes of the rocks changed as a result of the number of times they were shaken.

Rocks Table	
0 shakes	100 shakes
200 shakes	300 shakes
400 shakes	500 shakes
600 shakes	700 shakes

Draw Conclusions

4 **Drawing Conclusions** Why did the rocks change?

5 **Evaluating Results** How did the water change during the activity? Why did it change?

6 **Making Predictions** What would happen if you used a much harder rock, such as granite, for this experiment?

7 **Interpreting Information** How do the results of this experiment compare with what happens in a river?

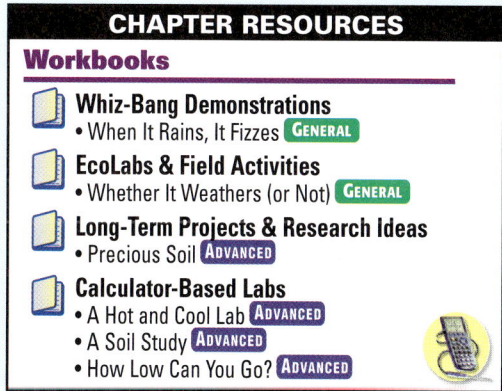

CHAPTER RESOURCES

Workbooks

Whiz-Bang Demonstrations
• When It Rains, It Fizzes GENERAL

EcoLabs & Field Activities
• Whether It Weathers (or Not) GENERAL

Long-Term Projects & Research Ideas
• Precious Soil ADVANCED

Calculator-Based Labs
• A Hot and Cool Lab ADVANCED
• A Soil Study ADVANCED
• How Low Can You Go? ADVANCED

Chapter Review

Assignment Guide

SECTION	QUESTIONS
1	1, 3, 5, 11, 17, 21, 24–26
2	6, 12, 18, 19, 21, 23
3	4, 7, 8, 13, 20
4	2, 9, 10, 14, 15, 16, 22

ANSWERS

Using Key Terms

1. Sample answer: Abrasion is the grinding and wearing away of rock surfaces through the mechanical action of other rock or sand particles. Soil texture is the soil quality that is based on the proportions of soil particles.

2. Sample answer: Soil conservation can help ensure that there will be enough fertile soil in which to plant crops. Erosion may occur if the soil is not covered with vegetation.

3. Sample answer: Mechanical weathering is the breaking down of rock by physical means. Chemical weathering is the process by which rocks break down as a result of chemical reactions.

4. Sample answer: Soil is a mixture of organic material, water, minerals and air that support the growth of vegetation. Parent rock is the rock formation that is the source of mineral fragments in the soil.

USING KEY TERMS

1. In your own words, write a definition for each of the following terms: *abrasion* and *soil texture*.

2. Use each of the following terms in a separate sentence: *soil conservation* and *erosion*.

For each pair of terms, explain how the meanings of the terms differ.

3. *mechanical weathering* and *chemical weathering*

4. *soil* and *parent rock*

UNDERSTANDING KEY IDEAS

Multiple Choice

5. Which of the following processes is a possible effect of water?
   a. mechanical weathering
   b. chemical weathering
   c. abrasion
   d. All of the above

6. In which climate would you find the fastest rate of chemical weathering?
   a. a warm, humid climate
   b. a cold, humid climate
   c. a cold, dry climate
   d. a warm, dry climate

7. Which of the following properties does soil texture affect?
   a. soil pH
   b. soil temperature
   c. soil consistency
   d. None of the above

8. Which of the following properties describes a soil's ability to supply nutrients?
   a. soil structure
   b. infiltration
   c. soil fertility
   d. consistency

9. Soil is important because it provides
   a. housing for animals.
   b. nutrients for plants.
   c. storage for water.
   d. All of the above

10. Which of the following soil conservation techniques prevents erosion?
    a. contour plowing
    b. terracing
    c. no-till farming
    d. All of the above

Short Answer

11. Describe the two major types of weathering.

12. Why is Devils Tower higher than the surrounding area?

13. Why is soil in temperate forests thick and fertile?

14. What can happen to soil when soil conservation is not practiced?

15. Describe the process of land degradation.

16. How do cover crops help prevent soil erosion?

Understanding Key Ideas

5. d	**8.** c
6. a	**9.** d
7. c	**10.** d

11. Mechanical weathering is the physical process of breaking down rock and minerals into smaller pieces. Chemical weathering is a chemical reaction that breaks down rock and minerals by chemical reactions.

12. The less resistant rock surrounding the original volcano weathered faster than the more resistant rock of Devils Tower.

13. Soil in temperate forests experiences high rates of weathering, which increases the rate of soil production.

14. Soil can be eroded or damaged if soil conservation is not practiced.

15. Overused soil loses its nutrients. Plants are unable to grow in this soil without nutrients. Without plants, water cannot be held and cycled, and the area can become a desert.

16. Cover crops protect the soil from wind, rain, and other agents of erosion.

17. Concept Mapping Use the following terms to create a concept map: *weathering, chemical weathering, mechanical weathering, abrasion, ice wedging, oxidation,* and *soil.*

18. Analyzing Processes Heat generally speeds up chemical reactions. But weathering, including chemical weathering, is usually slowest in hot, dry climates. Why?

19. Making Inferences Mechanical weathering, such as ice wedging, increases surface area by breaking larger rocks into smaller rocks. Draw conclusions about how mechanical weathering can affect the rate of chemical weathering.

20. Evaluating Data A scientist has a new theory. She believes that climates that receive heavy rains all year long have thin topsoil. Given what you have learned, decide if the scientist's theory is correct. Explain your answer.

21. Analyzing Processes What forms of mechanical and chemical weathering would be most common in the desert? Explain your answer.

22. Applying Concepts If you had to plant a crop on a steep hill, what soil conservation techniques would you use to prevent erosion?

23. Making Comparisons Compare the weathering processes in a warm, humid climate with those in a dry, cold climate.

INTERPRETING GRAPHICS

The graph below shows how the density of water changes when temperature changes. The denser a substance is, the less volume it occupies. In other words, as most substances get colder, they contract and become denser. But water is unlike most other substances. When water freezes, it expands and becomes less dense. Use the graph below to answer the questions that follow.

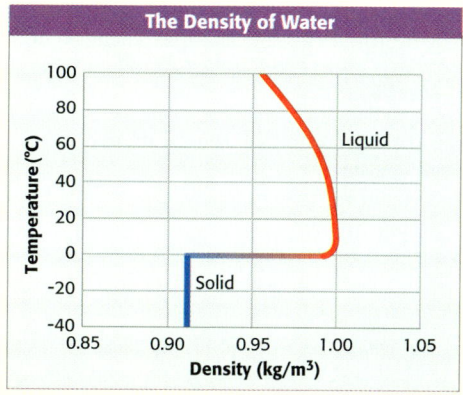

The Density of Water

24. Which has the greater density: water at 40°C or water at –20°C?

25. How would the line in the graph look if water behaved like most other liquids?

26. Which substance would be a more effective agent of mechanical weathering: water or another liquid? Why?

19. Mechanical weathering increases the surface area of rock, which exposes more of the rock to the effects of chemical weathering and increases the rate of weathering.

20. Sample answer: The scientist's theory is correct because heavy rain will leach the nutrients out of the soil, which will make the topsoil thin.

21. Abrasion and oxidation would probably be the most common in desert climates. Although the rate of oxidation and abrasion may increase with the presence of water, both processes may occur when little or no water is present.

22. Terracing and planting a cover crop between harvests would help prevent erosion on a steep slope.

23. Warm, humid climates have a higher rate of weathering than cold, dry climates do because the warmer temperature and moisture speed up chemical and mechanical weathering processes.

Interpreting Graphics

24. water at 40°C

25. The line would slope downward from left to right because most other liquids would have an increase in density as they became colder.

26. Water is more effective than other liquids because it expands when it freezes.

Critical Thinking

17. 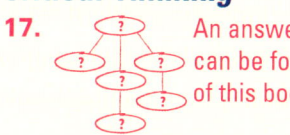 An answer to this exercise can be found at the end of this book.

18. Hot, dry climates generally have less precipitation than more temperate climates. Moisture enables chemical weathering to occur more quickly. The lack of moisture inhibits all processes of mechanical weathering except abrasion.

CHAPTER RESOURCES

Chapter Resource File
- Chapter Review **GENERAL**
- Chapter Test A **GENERAL**
- Chapter Test B **ADVANCED**
- Chapter Test C **SPECIAL NEEDS**
- Vocabulary Activity **GENERAL**

Workbooks

Study Guide
• Assessment resources are also available in Spanish.

Standardized Test Preparation

Answers to the standardized test preparation can help you identify student misconceptions and misunderstandings.

READING

Passage 1

1. B
2. F
3. C

Question 1: If students misinterpret the passage, they may think that castings have negative effect on plant growth. Therefore, they might conclude that the word *enhance* means "to weaken" or "to decrease." However, as it is used in the sentence, *enhance* means "to improve."

READING

Read each of the passages below. Then, answer the questions that follow each passage.

Passage 1 Earthworms are very important for forming soil. As they search for food by digging tunnels in the soil, they expose rocks and minerals to the effects of weathering. Over time, this process makes new soil. And as the worms dig tunnels, they mix the soil, which allows air and water and smaller organisms to move deeper into the soil. Worms have huge appetites. They eat organic matter and other materials in the soil. One earthworm can eat an amount equal to about half its body weight each day! Eating all of that food means that earthworms leave behind a lot of waste. Earthworm wastes, called *castings,* are very high in nutrients and make excellent natural fertilizer. Castings enrich the soil and <u>enhance</u> plant growth.

1. In the passage, what does *enhance* mean?
   - **A** to weaken
   - **B** to improve
   - **C** to smooth out
   - **D** to decrease

2. According to the passage, the earthworms
   - **F** eat organic matter and other materials in soil.
   - **G** do not have much of an appetite.
   - **H** love to eat castings.
   - **I** cannot digest organic matter in soil.

3. Which of the following statements is a fact according to the passage?
   - **A** Earthworms are not important for forming soil.
   - **B** Earthworms only eat organic matter in the soil.
   - **C** An earthworm can eat an amount that equals half its body weight each day.
   - **D** Earthworms eat little food but leave behind a lot of waste.

Passage 2 Worms are not the only living things that help create soil. Plants also play a part in the weathering process. As the roots of plants grow and seek out water and nutrients, they help break large rock fragments into smaller ones. Have you ever seen a plant growing in a crack in the sidewalk? As the plant grows, its roots spread into tiny cracks in the sidewalk. These roots apply pressure to the cracks, and over time, the cracks get bigger. As the plants make the cracks bigger, ice wedging can occur more readily. As the cracks expand, more water runs into them. When the water freezes, it expands and presses against the walls of the crack, which makes the crack even larger. Over time, the weathering caused by water, plants, and worms helps break down rock to form soil.

1. How do plants make it easier for ice wedging to occur?
   - **A** Plant roots block the cracks and don't allow water to enter.
   - **B** Plant roots provide moisture to cracks.
   - **C** Plant roots make the cracks larger, which allows more water to enter the cracks.
   - **D** Plants absorb excess water from cracks.

2. For ice wedging to occur,
   - **F** water in cracks must freeze.
   - **G** plant roots must widen cracks.
   - **H** acid is needed.
   - **I** water is not needed.

3. Which of the following statements is a fact according to the passage?
   - **A** Plant roots can strangle earthworms.
   - **B** Earthworms eat plant roots.
   - **C** Plant roots cannot crack sidewalks.
   - **D** Plant roots break large rock fragments into smaller ones.

Passage 2

1. C
2. F
3. D

 TEST DOCTOR

Question 2: The second half of the passage describes ice wedging: "When the water freezes, it expands and presses against the walls of the crack, which makes the crack even larger." However, because the beginning of the passage describes how the roots of plants can widen cracks, students may associate plants with ice wedging.

The graph below shows the average yearly rainfall in five locations. Use the graph below to answer the questions that follow.

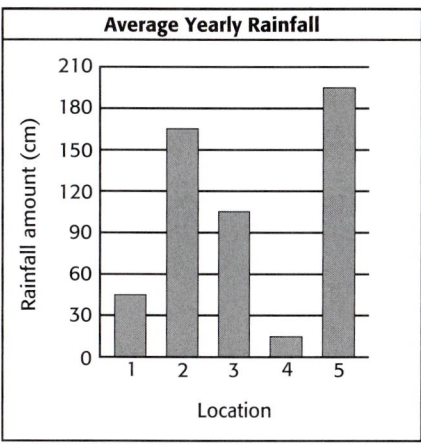

Average Yearly Rainfall

1. Which location has the **most** average yearly rainfall?
   - A 1
   - B 2
   - C 4
   - D 5

2. At which location would you expect to find the **most** chemical weathering?
   - F 1
   - G 3
   - H 4
   - I 5

3. At which location would you expect to find the **least** amount of chemical weathering?
   - A 2
   - B 3
   - C 4
   - D 5

Read each question below, and choose the best answer.

1. If an earthworm that weighs 1.5 g eats an amount equal to half its body weight in a day, how much does the earthworm eat in 1 week?
   - A 10.5 g
   - B 7 g
   - C 5.25 g
   - D 1.5 g

2. Calculate the surface area of a cube that measures 3 cm by 3 cm.
   - F 9 cm
   - G 9 cm^2
   - H 54 cm
   - I 54 cm^2

3. If a mountain peak weathers away 2 cm every 6 years, how many years will the mountain peak take to weather away 1 m?
   - A 8 years
   - B 12 years
   - C 180 years
   - D 300 years

4. The rock ledge that lies under a waterfall erodes about 3 cm each year. How much of the rock will erode over a period of 18 months?
   - F 4.5 cm
   - G 6 cm
   - H 21 cm
   - I 54 cm

5. A garden shop charges $0.30 for each ground-cover seedling. How many seedlings can you buy for $6.00?
   - A 5 seedlings
   - B 18 seedlings
   - C 20 seedlings
   - D 200 seedlings

Standardized Test Preparation

1. D
2. I
3. C

 **TEST DOCTOR**

Question 2: Water can increase the rate of chemical weathering. Students must know and apply this information to answer the question correctly.

1. C
2. I
3. D
4. F
5. C

 **TEST DOCTOR**

Question 2: The question asks for the surface area of the entire cube, not one side of the cube. After calculating the surface area of one side of the cube, students must then multiply by 6 to obtain the correct answer.

CHAPTER RESOURCES

Chapter Resource File

 • Standardized Test Preparation **GENERAL**

State Resources

 For specific resources for your state, visit **go.hrw.com** and type in the keyword **HSMSTR**.

Science, Technology, and Society

Background

Earth is not the only planet that experiences dust storms. NASA's *Mars Global Surveyor* and *Hubble Space Telescope* have given scientists amazing views of dust storms in the Martian atmosphere. Martian dust storms are far larger than those experienced on Earth. Martian dust storms can engulf the entire planet and last for months at a time. Students can find images of global dust storms online.

Scientific Discovery

Background

Brad Werner and Mark Kessler, the scientists who studied the soil patterns in Alaska and on the Norwegian Islands applied the principle of *self-organization* to their work. Self-organization looks for an explanation of change on a large scale. For example, it looks beyond the physics of a single grain of soil or an individual rock and looks at a system as a whole. According to the theory of self-organization, many phenomena on the Earth's surface, such as the soil patterns in Alaska, are responsible for their own development and maintenance over long periods of time.

Science in Action

Science, Technology, and Society

Flying Fertilizer

Would you believe that dust from storms in large deserts can be transported over the oceans to different continents? Dust from the Gobi Desert in China has traveled all the way to Hawaii! In many cases, the dust is a welcome guest. Iron in dust from the Sahara, a desert in Africa, fertilizes the canopies of South American rain forests. In fact, research has shown that the canopies of Central and South American rain forests get much of their nutrients from dust from the Sahara!

Social Studies ACTIVITY

Find pictures on the Internet or in magazines that show how people in rain forests live. Make a poster by using the pictures you find.

Scientific Discoveries

Strange Soil

Mysterious patterns of circles, polygons, and stripes were discovered in the soil in remote areas in Alaska and the Norwegian islands. At first, scientists were puzzled by these strange designs in remote areas. Then, the scientists discovered that these patterns were created by the area's weathering process, which includes cycles of freezing and thawing. When the soil freezes, the soil expands. When the soil thaws, the soil contracts. This process moves and sorts the particles of the soil into patterns.

Language Arts ACTIVITY

WRITING SKILL Write a creative short story describing what life would be like if you were a soil circle on one of these remote islands.

Answer to Social Studies Activity

Students' posters should include adaptive activities specific to the people who live in a tropical rain forest. For example, a student may include a picture of Fijians building a traditional house with high ceilings that allows the air to circulate in the house such that the air won't become stuffy in warm, humid climates.

Answer to Language Arts Activity

Students' stories should include a description of the freezing and thawing climate on the remote island. Encourage students to conduct more research about the climate, plants, and animals of these islands.

J. David Bamberger

Habitat Restoration J. David Bamberger knows how important taking care of the environment is. Therefore, he has turned his ranch into the largest habitat restoration project in Texas. For Bamberger, restoring the habitat started with restoring the soil. One way Bamberger restored the soil was to manage the grazing of the grasslands and to make sure that grazing animals didn't expose the soil. Overgrazing causes soil erosion. When cattle clear the land of its grasses, the soil is exposed to wind and rain, which can wash the topsoil away.

Bamberger also cleared his land of most of the shrub, *juniper*. Juniper requires so much water per day that it leaves little water in the soil for the grasses and wildflowers. The change in the ranch since Bamberger first bought it in 1959 is most obvious at the fence-line border of his ranch. Beyond the fence is a small forest of junipers and little other vegetation. On Bamberger's side, the ranch is lush with grasses, wildflowers, trees, and shrubs.

Math ACTIVITY

Bamberger's ranch is 2,300 hectares. There are 0.405 hectares in 1 acre. How many acres is Bamberger's ranch?

go.hrw.com

To learn more about these Science in Action topics, visit **go.hrw.com** and type in the keyword **HZ5WSFF.**

Current Science

Check out Current Science® articles related to this chapter by visiting **go.hrw.com.** Just type in the keyword **HZ5CS10.**

Answer to Math Activity
2,300 hectares ÷ 0.405 hectares/acre = 5,700 acres

Agents of Erosion and Deposition
Chapter Planning Guide

Compression guide:
To shorten instruction because of time limitations, omit the Chapter Lab.

OBJECTIVES	LABS, DEMONSTRATIONS, AND ACTIVITIES	TECHNOLOGY RESOURCES
PACING • 90 min pp. 60–67 **Chapter Opener**	**SE** Start-up Activity, p. 61 ◆ `GENERAL`	**OSP** Parent Letter ■ `GENERAL` **CD** Student Edition on CD-ROM **CD** Guided Reading Audio CD ■ **TR** Chapter Starter Transparency* **VID** Brain Food Video Quiz
Section 1 Shoreline Erosion and Deposition • Explain how energy from waves affects a shoreline. • Identify six shoreline features created by wave erosion. • Explain how wave deposits form beaches. • Describe how sand moves along a beach.	**TE** Activity Illustrating Beach Erosion, p. 62 `BASIC` **TE** Connection Activity Math, p. 63 `BASIC` **TE** Group Activity Coastal Features Board Game, p. 64 `GENERAL` **LB** Whiz-Bang Demonstrations Between a Rock and a Hard Place* ◆ `BASIC` **LB** Whiz-Bang Demonstrations Rising Mountains* ◆ `GENERAL`	**CRF** Lesson Plans* **TR** Bellringer Transparency* **TR** Coastal Landforms Created by Wave Erosion A* **TR** Coastal Landforms Created by Wave Erosion B* **CD** Science Tutor
PACING • 45 min pp. 68–71 **Section 2 Wind Erosion and Deposition** • Explain why some areas are more affected by wind erosion than other areas are. • Describe the process of saltation. • Identify three landforms that result from wind erosion and deposition. • Explain how dunes move.	**SE** Quick Lab Making Desert Pavement, p. 69 ◆ `GENERAL` **CRF** Datasheet for Quick Lab* **SE** Connection to Language Arts The Dust Bowl, p. 70 `GENERAL` **TE** Connection Activity Life Science, p. 70 `BASIC` **SE** Model-Making Lab Dune Movement, p. 95 `GENERAL` **CRF** Datasheet for LabBook*	**CRF** Lesson Plans* **TR** Bellringer Transparency* **CD** Science Tutor
PACING • 90 min pp. 72–77 **Section 3 Erosion and Deposition by Ice** • Explain the difference between alpine glaciers and continental glaciers. • Describe two ways in which glaciers move. • Identify five landscape features formed by alpine glaciers. • Identify four types of moraines.	**TE** Group Activity No Glaciers?, p. 72 `GENERAL` **SE** School-to-Home Activity The *Titanic*, p. 73 `GENERAL` **TE** Activity Describing Glacier Formation, p. 73 `BASIC` **TE** Demonstration Glacier Movement, p. 73 ◆ `BASIC` **TE** Connection Activity Environmental Science, p. 74 `GENERAL` **TE** Group Activity Making Models, p. 75 `GENERAL` **TE** Connection Activity Language Arts, p. 75 `ADVANCED` **SE** Model-Making Lab Gliding Glaciers, p. 82 ◆ `GENERAL` **CRF** Datasheet for Chapter Lab* **SE** Skills Practice Lab Creating a Kettle, p. 96 `GENERAL` **CRF** Datasheet for LabBook*	**CRF** Lesson Plans* **TR** Bellringer Transparency* **TR** Landscape Features Carved by Alpine Glaciers* **VID** Lab Videos for Earth Science **CD** Science Tutor
PACING • 45 min pp. 78–81 **Section 4 The Effect of Gravity on Erosion and Deposition** • Explain the role of gravity as an agent of erosion and deposition. • Explain how angle of repose is related to mass movement • Describe four types of rapid mass movement. • Describe three factors that affect creep.	**TE** Activity Demonstrating Mass Movement, p. 79 ◆ `BASIC` **TE** Connection Activity Environmental Science, p. 79 `GENERAL` **LB** Long-Term Projects & Research Ideas Deep in the Mud* `ADVANCED` **SE** Science in Action Math, Social Studies, and Language Arts Activities, pp. 88–89 `GENERAL`	**CRF** Lesson Plans* **TR** Bellringer Transparency* **TR** LINK TO PHYSICAL SCIENCE Gravitational Force Depends on Mass* **TE** Internet Activity, p. 81 `GENERAL` **CRF** SciLinks Activity* `GENERAL` **CD** Science Tutor

PACING • 90 min

CHAPTER REVIEW, ASSESSMENT, AND STANDARDIZED TEST PREPARATION

CRF Vocabulary Activity* `GENERAL`
SE Chapter Review, pp. 84–85 `GENERAL`
CRF Chapter Review* ■ `GENERAL`
CRF Chapter Tests A* ■ `GENERAL`, B* `ADVANCED`, C* `SPECIAL NEEDS`
SE Standardized Test Preparation, pp. 86–87 `GENERAL`
CRF Standardized Test Preparation* `GENERAL`
CRF Performance-Based Assessment* `GENERAL`
OSP Test Generator `GENERAL`
CRF Test Item Listing* `GENERAL`

Online and Technology Resources

Visit **go.hrw.com** for a variety of free resources related to this textbook. Enter the keyword **HZ5ICE**.

Holt Online Learning

Students can access interactive problem-solving help and active visual concept development with the *Holt Science and Technology* Online Edition available at **www.hrw.com**.

 Guided Reading Audio CD
Also in Spanish

A direct reading of each chapter for auditory learners, reluctant readers, and Spanish-speaking students.

 Science Tutor
CD-ROM

Excellent for remediation and test practice.

SKILLS DEVELOPMENT RESOURCES	SECTION REVIEW AND ASSESSMENT	CORRELATIONS
SE Pre-Reading Activity, p. 60 `GENERAL` **OSP** Science Puzzlers, Twisters & Teasers `GENERAL`		National Science Education Standards SAI 1; ES 2a
CRF Directed Reading A* ■ `BASIC`, B* `SPECIAL NEEDS` **CRF** Vocabulary and Section Summary* ■ `GENERAL` **SE** Reading Strategy Reading Organizer, p. 62 `GENERAL` **SE** Math Practice Counting Waves, p. 63 `GENERAL` **MS** Math Skills for Science The Unit Factor and Dimensional Analysis* `GENERAL` **SS** Science Skills Using Your Senses* `GENERAL`	**SE** Reading Checks, pp. 63, 65, 66 `GENERAL` **TE** Reteaching, p. 66 `BASIC` **TE** Quiz, p. 66 `GENERAL` **TE** Alternative Assessment, p. 66 `GENERAL` **SE** Section Review,* p. 67 ■ `GENERAL` **CRF** Section Quiz* ■ `GENERAL`	UCP 3, SAI 1, SPSP 2, 3, ES 1c
CRF Directed Reading A* ■ `BASIC`, B* `SPECIAL NEEDS` **CRF** Vocabulary and Section Summary* ■ `GENERAL` **SE** Reading Strategy Reading Organizer, p. 68 `GENERAL` **TE** Inclusion Strategies, p. 69 ◆ **SE** Connection to Language Arts The Dust Bowl, p. 70 `GENERAL` **CRF** Critical Thinking A Future in Sand* `ADVANCED`	**SE** Reading Checks, pp. 69, 71 `GENERAL` **TE** Reteaching, p. 70 ◆ `BASIC` **TE** Quiz, p. 70 `GENERAL` **TE** Alternative Assessment, p. 70 `GENERAL` **SE** Section Review,* p. 71 ■ `GENERAL` **CRF** Section Quiz* ■ `GENERAL`	SAI 1; ST 2; SPSP 2; HNS 1; ES 1c, 2a; *LabBook:* UCP 2, 3; SAI 1
CRF Directed Reading A* ■ `BASIC`, B* `SPECIAL NEEDS` **CRF** Vocabulary and Section Summary* ■ `GENERAL` **SE** Reading Strategy Discussion, p. 72 `GENERAL` **SE** Math Practice Speed of a Glacier, p. 74 `GENERAL` **TE** Reading Strategy Preparing Tables, p. 75 `BASIC` **TE** Inclusion Strategies, p. 76 ◆ **MS** Math Skills for Science Using Proportions and Cross-Multiplication* `GENERAL` **CRF** Reinforcement Worksheet An Alpine Vacation* `BASIC`	**SE** Reading Checks, pp. 72, 77 `GENERAL` **TE** Reteaching, p. 76 `BASIC` **TE** Quiz, p. 76 `GENERAL` **TE** Alternative Assessment, p. 76 `GENERAL` **SE** Section Review,* p. 77 ■ `GENERAL` **CRF** Section Quiz* ■ `GENERAL`	UCP 2, 3; SAI 1; SPSP 3; ES 1c, 2a; *Chapter Lab:* UCP 2; SAI 1; ST 1; *LabBook:* UCP 2, SAI 1, ES 1c
CRF Directed Reading A* ■ `BASIC`, B* `SPECIAL NEEDS` **CRF** Vocabulary and Section Summary* ■ `GENERAL` **SE** Reading Strategy Prediction Guide, p. 78 `GENERAL`	**SE** Reading Checks, pp. 79, 80 `GENERAL` **TE** Reteaching, p. 80 `BASIC` **TE** Quiz, p. 80 `GENERAL` **TE** Alternative Assessment, p. 80 `GENERAL` **SE** Section Review,* p. 81 ■ `GENERAL` **CRF** Section Quiz* ■ `GENERAL`	SAI 1; SPSP 3, 4; ES 1c

One-Stop Planner® CD-ROM

This CD-ROM package includes:
- Lab Materials QuickList Software
- Holt Calendar Planner
- Customizable Lesson Plans
- Printable Worksheets
- ExamView® Test Generator
- Interactive Teacher Edition
- Holt PuzzlePro® Resources
- Holt PowerPoint® Resources

SCILINKS® NSTA

www.scilinks.org

Maintained by the **National Science Teachers Association.** See Chapter Enrichment pages for a complete list of topics.

Current Science®

Check out *Current Science* articles and activities by visiting the HRW Web site at go.hrw.com. Just type in the keyword **HZ5CS12T.**

Classroom Videos

- **Lab Videos** demonstrate the chapter lab.
- **Brain Food Video Quizzes** help students review the chapter material.
- **CNN Videos** bring science into your students' daily life.

Visual Resources

CHAPTER STARTER TRANSPARENCY

This Really Happened!

BELLRINGER TRANSPARENCIES

Section: Shoreline Erosion and Deposition
Where does sand come from? Many people find the sound of waves on a beach very relaxing and peaceful. However, each wave that comes ashore carries a certain amount of destructive force. Write a short poem about how ocean waves create sand from rock.

Write your poem in your **science journal.**

Section: Wind Erosion and Deposition
What causes wind? Write your answer in your **science journal.**

TEACHING TRANSPARENCIES

Coastal Landforms Created by Wave Erosion: A

Coastal Landforms Created by Wave Erosion: B

TEACHING TRANSPARENCIES

Landscape Features Carved by Alpine Glaciers

Gravitational Force Depends on Mass

Gravitational Force Depends on Distance

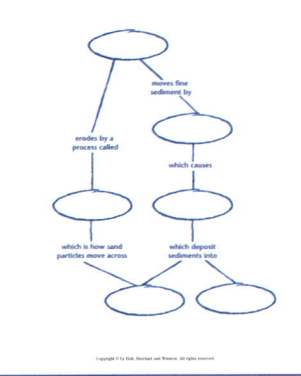

LINK TO PHYSICAL SCIENCE

Chapter: Matter and Motion

CONCEPT MAPPING TRANSPARENCY

Use the following terms to complete the concept map below:
dust storms, wind, saltation, deflation, loess, dunes

Planning Resources

LESSON PLANS

Lesson Plan   **SAMPLE**

Section: Waves

Pacing
Regular Schedule: with lab(s):2 days   without lab(s)if days
Block Schedule: with lab(s): 1 1/2 days without lab(s)if day

Objectives
1. Relate the seven properties of life to a living organism.
2. Describe seven themes that can help you to organize what you learn about biology.
3. Identify the tiny structures that make up all living organisms.
4. Differentiate between reproduction and heredity and between metabolism and homeostasis.

National Science Education Standards Covered
LSInter4:Cells have particular structures that underlie their functions.
LSMat1:Most cell functions involve chemical reactions.
LSBeh1:Cells store and use information to guide their functions.
UCP1:Cell functions are regulated.
SE1: Cells can differentiate and form complete multicellular organisms.
PS1: Species evolve over time.
ESS1: The great diversity of organisms is the result of more than 3.5 billion years of evolution.
ESS2: Natural selection and its evolutionary consequences provide a scientific explanation for the fossil record of ancient life forms as well as for the striking molecular similarities observed among the diverse species of living organisms.
ST1: The millions of different species of plants, animals, and microorganisms that live on Earth today are related by descent from common ancestors.
ST2: The energy for life primarily comes from the sun.
SPSP1: The complexity and organization of organisms accommodate the need for obtaining, transforming, transporting, releasing, and eliminating the matter and energy used to sustain the organism.
SPSP6: As matter and energy flows through different levels of organization of living systems—cells, organs, communities—and between living systems and the physical environment, chemical elements are recombined in different ways.
HNS1: Organisms have behavioral responses to internal changes and to external stimuli.

PARENT LETTER

SAMPLE

Dear Parent,

Your son's or daughter's science class will soon begin exploring the chapter entitled "The World of Physical Science." In this chapter, students will learn how the scientific method applies to the world of physical science and the role of physical science in the world. By the end of the chapter, students should demonstrate a clear understanding of the chapter's main ideas and be able to discuss the following topics:

1. physical science as the study of energy and matter (Section 1)
2. the role of physical science in the world around them (Section 1)
3. careers that rely on physical science (Section 1)
4. the steps used in the scientific method (Section 2)
5. examples of technology (Section 2)
6. how the scientific method is used to answer questions and solve problems (Section 2)
7. how our knowledge of science changes over time (Section 2)
8. how models represent real objects or systems (Section 3)
9. examples of different ways models are used in science (Section 3)
10. the importance of the International System of Units (Section 4)
11. the appropriate units to use for particular measurements (Section 4)
12. how area and density are derived quantities (Section 4)

Questions to Ask Along the Way

You can help your son or daughter learn about these topics by asking interesting questions such as the following:

• What are some surprising careers that use physical science?
• What is a characteristic of a good hypothesis?
• When is it a good idea to use a model?
• Why do Americans measure things in terms of inches and yards and meters ?

ALSO IN SPANISH

TEST ITEM LISTING

TEST ITEM LISTING
The World of Earth Science   **SAMPLE**

MULTIPLE CHOICE

One-Stop Planner® CD-ROM

This CD-ROM includes all of the resources shown here and the following time-saving tools:

• *Lab Materials QuickList Software*
• *Customizable lesson plans*
• *Holt Calendar Planner*
• *The powerful ExamView® Test Generator*

Meeting Individual Needs

DIRECTED READING A
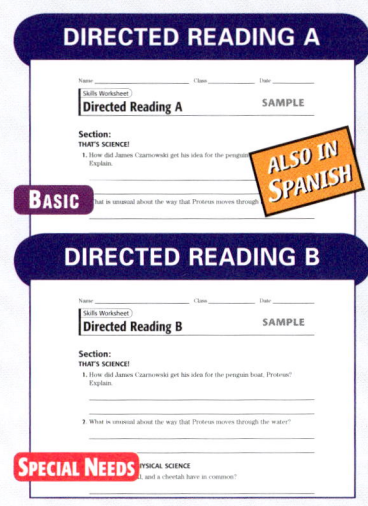
BASIC — ALSO IN SPANISH

DIRECTED READING B
SPECIAL NEEDS

VOCABULARY ACTIVITY
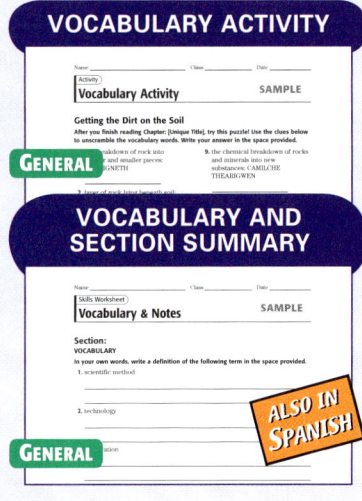
GENERAL

VOCABULARY AND SECTION SUMMARY
GENERAL — ALSO IN SPANISH

REINFORCEMENT
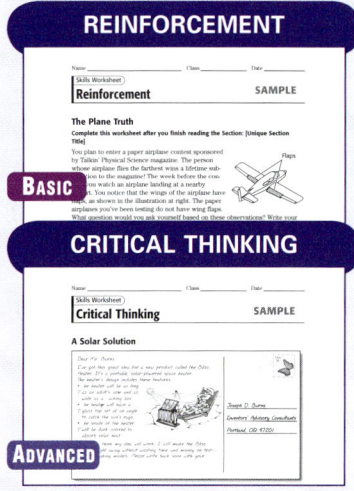
BASIC

CRITICAL THINKING
ADVANCED

SCILINKS ACTIVITY
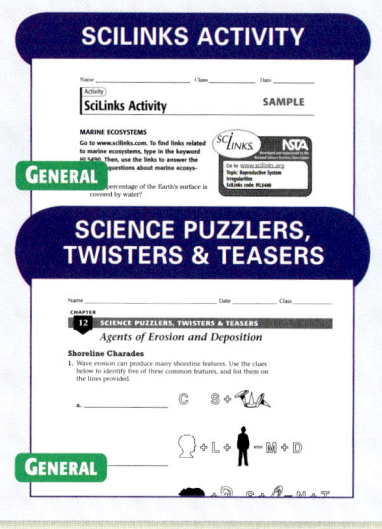
GENERAL

SCIENCE PUZZLERS, TWISTERS & TEASERS
GENERAL

Labs and Activities

LONG-TERM PROJECTS & RESEARCH IDEAS
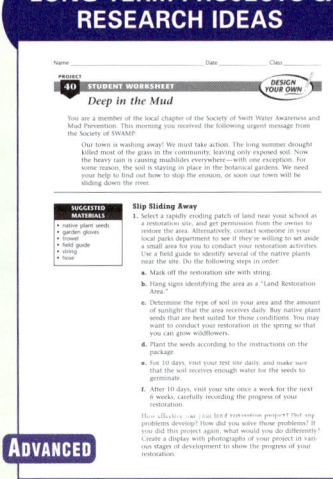
ADVANCED

WHIZ-BANG DEMONSTRATIONS
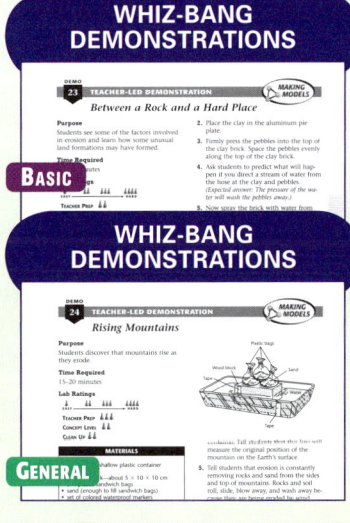
BASIC

WHIZ-BANG DEMONSTRATIONS
GENERAL

DATASHEETS FOR QUICKLABS
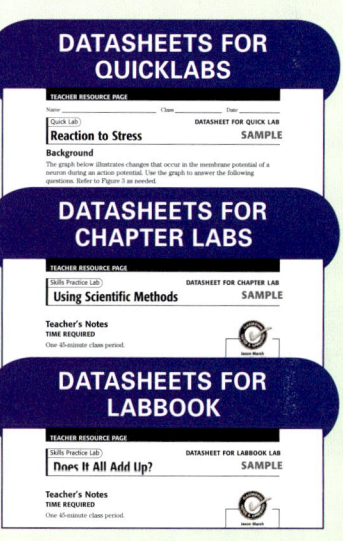

DATASHEETS FOR CHAPTER LABS

DATASHEETS FOR LABBOOK

Review and Assessments

SECTION QUIZ
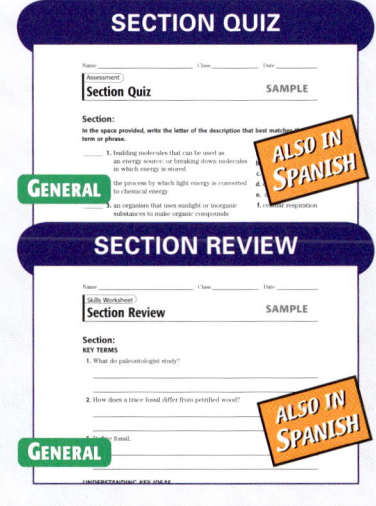
GENERAL — ALSO IN SPANISH

SECTION REVIEW
GENERAL — ALSO IN SPANISH

CHAPTER REVIEW
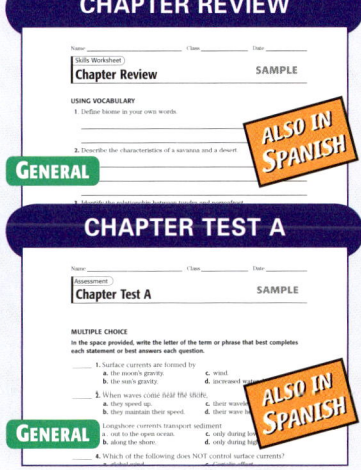
GENERAL — ALSO IN SPANISH

CHAPTER TEST A
GENERAL — ALSO IN SPANISH

CHAPTER TEST B
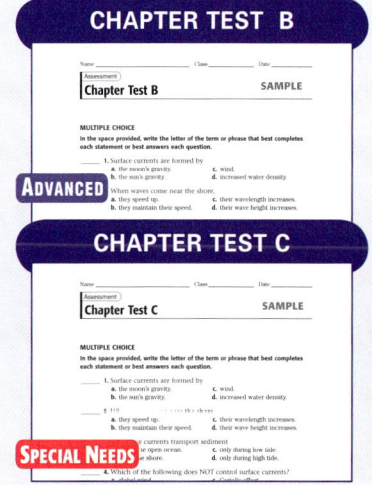
ADVANCED

CHAPTER TEST C
SPECIAL NEEDS

STANDARDIZED TEST PREPARATION
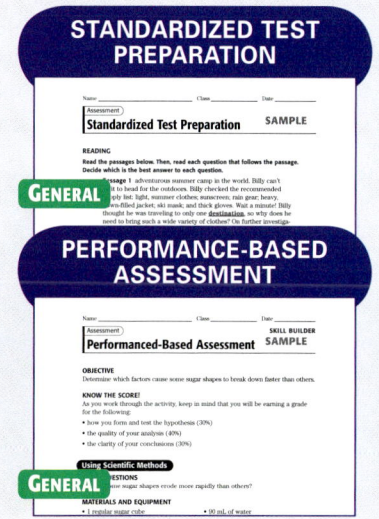
GENERAL

PERFORMANCE-BASED ASSESSMENT
GENERAL

This Chapter Enrichment provides relevant and interesting information to expand and enhance your presentation of the chapter material.

Section 1

Shoreline Erosion and Deposition

Acrobatic Waves

● To understand shoreline erosion, students may find it helpful to understand the forces acting in a breaking wave. Breaking waves can be thought of as somersaulting water. As waves move toward shallow coastal waters, the wavelengths shorten, the crests crowd together, and the wave heights grow. When a wave becomes too top heavy, it falls forward and rushes onto the shore. As the water flows back into the ocean, it carries sand and sediment with it.

The Origins of Cape Cod

● At the end of the last glacial period—10,000 years ago—glaciers receding across North America helped form Cape Cod, Massachusetts. Cape Cod was initially mounds of outwash, or debris left behind by the glaciers. These mounds of debris were then surrounded by the rising sea. Over time, currents eroded land and filled in depressions between the island mounds. Sandbars connected the islands to each other and to the mainland. Since its formation, Cape Cod has lost 3.2 km of coastline because of ocean erosion.

Section 2

Wind Erosion and Deposition

The Dust Bowl

● The Dust Bowl was a section of the Great Plains of the United States that extended from southeastern Colorado and southwestern Kansas to the panhandles of Texas and Oklahoma and to northeastern New Mexico. In the early 1930s, following years of overcultivation, the region

suffered a severe drought. Exposed topsoil was carried away by strong spring winds. Windblown soil sometimes blocked out the sun, and the dirt piled up in drifts like snow. Occasionally, huge dust storms blew across the country and reached the East Coast. The wind erosion was gradually halted when the federal government planted windbreaks, and large areas of grasslands were restored. The area had mostly recovered by the early 1940s.

Lost Cities of the Takla Makan Desert

● The Takla Makan desert in China's arid northwest is so inhospitable that its name in the local language means "Go in, and you don't come out." The desert is covered with treacherous dunes of fine, dry sand. Buried under those dunes are the remains of cities that prospered along the ancient Silk Road. The Silk Road was a trade route that connected China to civilizations in the West. NASA's Spaceborne Imaging Radar (SIR-C), which flew on space shuttles twice in 1994, is being used to examine the desert. The radar-imaging technology has already helped archeologists locate some cities and promises to help them find other ruins.

Section 3

Erosion and Deposition by Ice

Glaciers and Drinking Water

- Arapaho Glacier, a small, perennial ice sheet in Colorado, provides water to more than 75,000 people living in Boulder, Colorado. Many countries have explored the possibility of obtaining drinking water from glaciers. Some countries have even proposed towing icebergs into their harbors!

Is That a Fact!

- ◆ Glaciers flow at different rates. Most glaciers flow at an average rate of 20 cm/day or less, but some flow much faster. In 1936, the Black Rapids Glacier, in Alaska, was measured to flow at a rate of 30 m/day.

- ◆ If all of Earth's glaciers simultaneously melted, global sea levels would rise more than 65 m and would submerge coastal cities all over the world.

Battles on Siachen Glacier

- At 70 km long, the Siachen Glacier is one of the world's longest glaciers. The glacier is in the Karakoram Range, on the India-Pakistan border. The Siachen Glacier is also the site of the world's highest battles. Indian and Pakistani soldiers have fought over this disputed territory of Kashmir on peaks as high as 6,400 m (almost 21,000 ft).

The Great Lakes

- The Great Lakes were formed by the movement of ice sheets during the Pleistocene epoch. These glaciers advanced over the land and gouged out a series of deep basins. As the glaciers melted, the basins filled with meltwater, and the five Great Lakes were formed.

Section 4

The Effect of Gravity on Erosion and Deposition

Scree

- Stones and boulders loosened by weathering and carried downward by gravity may be deposited in long, loose heaps called *scree* at the base of a mountain.

Is That a Fact!

- ◆ When an earthquake that measured 5.0 on the Richter scale occurred near Mount St. Helens on May 18, 1980, it triggered a landslide of more than 2 km³ of rock and ice. Immediately afterward, the eruption of Mount St. Helens began, and an explosion of steam and volcanic gases produced a lahar that raced down the mountain at speeds of up to 250 km/h.

SciLinks is maintained by the National Science Teachers Association to provide you and your students with interesting, up-to-date links that will enrich your classroom presentation of the chapter.

Visit www.scilinks.org and enter the SciLinks code for more information about the topic listed.

Topic: Wave Erosion
SciLinks code: HSM1638

Topic: Wind Erosion
SciLinks code: HSM1669

Topic: Glaciers
SciLinks code: HSM0675

Topic: Mass Movements
SciLinks code: HSM6295

Overview

Tell students that this chapter will help them learn about the processes of erosion and deposition by water, wind, ice, and gravity.

Assessing Prior Knowledge

Students should be familiar with the following topics:

• weathering
• soil formation

Identifying Misconceptions

Some students may be confused about the way that waves move. You may want to explain to students that as wave energy moves through water, the water itself does not travel very far toward the shore. When a wave passes through water, the water moves up and down. It is only when waves reach the shore and wave height exceeds the depth of the water, that water tumbles forward. In addition, students may not realize how much of the United States was shaped by glacial erosion. Use a map to show students how much of the northern United States was covered by glaciers during the last glacial period.

Agents of Erosion and Deposition

About the PHOTO

The results of erosion can often be dramatic. For example, this sinkhole formed in a parking lot in Atlanta, Georgia, when water running underground eventually caused the surface of the land to collapse.

PRE-READING ACTIVITY

FOLDNOTES **Layered Book** Before you read the chapter, create the FoldNote entitled "The Layered Book" described in the **Study Skills** section of the Appendix. Label the tabs of the layered book with "Shoreline erosion and deposition," "Wind erosion and deposition," and "Erosion and deposition by ice." As you read the chapter, write information you learn about each category under the appropriate tab.

Standards Correlations

National Science Education Standards

The following codes indicate the National Science Education Standards that correlate to this chapter. The full text of the standards is at the front of the book.

Chapter Opener
SAI 1; ES 2a

Section 1 Shoreline Erosion and Deposition
UCP 3, SAI 1, SPSP 2, 3, ES 1c

Section 2 Wind Erosion and Deposition
SAI 1; ST 2; SPSP 2; HNS 1; ES 1c, 2a; *LabBook:* UCP 2, 3; SAI 1

Section 3 Erosion and Deposition by Ice
UCP 2, 3; SAI 1; SPSP 3; ES 1c, 2a; *LabBook:* UCP 2, SAI 1, ES 1c

Section 4 The Effect of Gravity on Erosion and Deposition
SAI 1; SPSP 3, 4; ES 1c

Chapter Lab
UCP 2; SAI 1; ST 1

START-UP ACTIVITY

MATERIALS

FOR EACH GROUP
• block, wooden or plastic
• sand
• washtub
• water

Answers

1. Sample answer: The shoreline is slowly receding, or eroding.

2. Small waves erode less shoreline than large waves do. Therefore, large waves have a greater impact on the shoreline.

START-UP ACTIVITY

Making Waves

Above ground or below, water plays an important role in the erosion and deposition of rock and soil. A shoreline is a good example of how water shapes the Earth's surface by erosion and deposition. Did you know that shorelines are shaped by crashing waves? Build a model shoreline, and see for yourself!

Procedure

1. Make a shoreline by adding **sand** to one end of a **washtub**. Fill the washtub with **water** to a depth of 5 cm. Sketch the shoreline profile (side view), and label it "A."

2. Place a **block** at the end of the washtub opposite the beach.

3. Move the block up and down very slowly to create small waves for 2 min. Sketch the new shoreline profile, and label it "B."

4. Now, move the block up and down more rapidly to create large waves for 2 min. Sketch the new shoreline profile, and label it "C."

Analysis

1. Compare the three shoreline profiles. What is happening to the shoreline?

2. How do small waves and large waves erode the shoreline differently?

Chapter Review
UCP 2, 3; SAI 1; SPSP 3, 4, 5; ES 1c, 2a

Science in Action
SPSP 3; HNS 1, 3

This Really Happened!

The waves struck at the ocean cliffs, releasing their energy as they do every day. But on this day the waves seemed different—they were larger and more explosive, cutting away at the rock with each crash.

This is all part of the ongoing natural process of coastal erosion along the California shoreline and similar shorelines throughout the world. Winter storms create powerful waves that crash into the cliffs, breaking off pieces of rock that fall

Chapter Starter Transparency
Use this transparency to help students begin thinking about the erosive force of breaking waves.

CHAPTER RESOURCES

Technology

 Transparencies
• Chapter Starter Transparency

READING SKILLS

Student Edition on CD-ROM

Guided Reading Audio CD
• English or Spanish

Classroom Videos
• Brain Food Video Quiz

Workbooks

Science Puzzlers, Twisters & Teasers
• Agents of Erosion and Deposition **GENERAL**

Focus

Overview

This section explores how wave erosion and deposition shape shorelines. Students explore coastal landforms created by wave erosion, such as sea cliffs, sea stacks, sea arches, sea caves, headlands, and wave-cut terraces. Students also learn about the formation of beaches and offshore landforms by deposition.

Bellringer

Writing Ask students to think about where sand comes from. Have them write a short poem about how ocean waves create sand from rock.

Motivate

 ——————— **GENERAL**

Illustrating Beach Erosion

Explain to students that shorelines are dynamic, changing environments because ocean waves and currents continually erode and redeposit sand. Have each student draw a "filmstrip" illustrating the changes that could occur in the history of a beach. Have students illustrate the processes that cause these changes and write a caption that explains each frame of the film strip. **English Language Learners**

LS Visual

READING WARM-UP

Objectives

- Explain how energy from waves affects a shoreline.
- Identify six shoreline features created by wave erosion.
- Explain how wave deposits form beaches.
- Describe how sand moves along a beach.

Terms to Learn
shoreline
beach

READING STRATEGY

Reading Organizer As you read this section, create an outline of the section. Use the headings from the section in your outline.

Shoreline Erosion and Deposition

Think about the last time you were at a beach. Where did all of the sand come from?

Two basic ingredients are necessary to make sand: rock and energy. The rock is usually available on the shore. The energy is provided by waves that travel through water. When waves crash into rocks over long periods of time, the rocks are broken down into smaller and smaller pieces until they become sand.

As you read on, you will learn how wave erosion and deposition shape the shoreline. A **shoreline** is simply the place where land and a body of water meet. Waves usually play a major role in building up and breaking down the shoreline.

Wave Energy

As the wind moves across the ocean surface, it produces ripples called *waves*. The size of a wave depends on how hard the wind is blowing and how long the wind blows. The harder and longer the wind blows, the bigger the wave.

The wind that results from summer hurricanes and severe winter storms produces large waves that cause dramatic shoreline erosion. Waves may travel hundreds or even thousands of kilometers from a storm before reaching the shoreline. Some of the largest waves to reach the California coast are produced by storms as far away as Australia. So, the California surfer in **Figure 1** can ride a wave that formed on the other side of the Pacific Ocean!

Figure 1 *Waves produced by storms on the other side of the Pacific Ocean propel this surfer toward a California shore.*

CHAPTER RESOURCES

Chapter Resource File

- **Lesson Plan**
- **Directed Reading A** BASIC
- **Directed Reading B** SPECIAL NEEDS

Technology

- **Transparencies**
  - Bellringer

Workbooks

- **Math Skills for Science**
  - The Unit Factor and Dimensional Analysis GENERAL

MISCONCEPTION ALERT

Wave Movement A popular misconception is that a wave is a moving wall of water. Water actually moves up and down rather than forward as wave energy travels through it. When waves break, however, they do carry water with them.

Wave Trains

When you drop a pebble into a pond, is there just one ripple? Of course not. Waves, like ripples, don't move alone. As shown in **Figure 2,** waves travel in groups called *wave trains*. As wave trains move away from their source, they travel through the ocean water uninterrupted. But when waves reach shallow water, the bottom of the wave drags against the sea floor, slowing the wave down. The upper part of the wave moves more rapidly and grows taller. When the top of the wave becomes so tall that it cannot support itself, it begins to curl and break. These breaking waves are known as *surf*. Now you know how surfers got their name. The *wave period* is the time interval between breaking waves. Wave periods are usually 10 to 20 s long.

The Pounding Surf

Look at **Figure 3,** and you will get an idea of how sand is made. A tremendous amount of energy is released when waves break. A crashing wave can break solid rock and throw broken rocks back against the shore. As the rushing water in breaking waves enters cracks in rock, it helps break off large boulders and wash away fine grains of sand. The loose sand picked up by waves wears down and polishes coastal rocks. As a result of these actions, rock is broken down into smaller and smaller pieces that eventually become sand.

✓ Reading Check How do waves help break down rock into sand? (*See the Appendix for answers to Reading Checks.*)

Figure 2 *Because waves travel in wave trains, they break at regular intervals.*

shoreline the boundary between land and a body of water

MATH PRACTICE

Counting Waves

If the wave period is 10 s, approximately how many waves reach a shoreline in a day? (Hint: Calculate how many waves occur in an hour, and multiply that number by the number of hours in a day.)

Figure 3 *Breaking waves crash against the rocky shore, releasing their energy.*

Coastal Features Board Game
To reinforce section concepts, divide the class into small groups, and challenge each group to create a board game. Tell students that the object of the game is for players to visit as many coastal landforms as they can. Provide each group with poster board, plain index cards, and markers. Direct groups to create a game board that leads players along a "coastline" and allows them to encounter the features they have learned about. Have students use the index cards to write questions and clues to direct players' movements along the coast. For example, students might write, "If you can describe how a sea arch forms, you may move ahead to the sea stack. If not, you lose a turn." Have groups create written game rules, exchange games, and play their games. (**Note:** It might be helpful to brainstorm chapter concepts on the board before students create their games.) **LS** Logical/Kinesthetic **English Language Learners**

Wave Erosion

Wave erosion produces a variety of features along a shoreline. *Sea cliffs* are formed when waves erode and undercut rock to produce steep slopes. Waves strike the base of the cliff, which wears away the soil and rock and makes the cliff steeper. The rate at which the sea cliffs erode depends on the hardness of the rock and the energy of the waves. Sea cliffs made of hard rock, such as granite, erode very slowly. Sea cliffs made of soft rock, such as shale, erode more rapidly, especially during storms.

Figure 4 Coastal Landforms Created by Wave Erosion

Sea stacks are offshore columns of resistant rock that were once connected to the mainland. In these instances, waves have eroded the mainland, leaving behind isolated columns of rock.

Sea arches form when wave action continues to erode a sea cave, cutting completely through the rock.

Sea caves form when waves cut large holes into fractured or weak rock along the base of sea cliffs. Sea caves are common in cliffs composed of sedimentary rock.

Cultural Awareness

Polynesian Cultures The Polynesians are considered to have been some of the greatest navigators of the ancient world. Polynesians visited and inhabited more than 10,000 islands throughout the South Pacific. They navigated not by using maps but by carefully observing stars, winds, and waves. On cloudy nights, they listened to the way the waves rocked and slapped against their dugout canoes. The Polynesians understood how wave patterns could indicate the direction of land or the presence of dangerous reefs or sandbars. Encourage students to discover more about Polynesian cultures of the past and present.

Shaping a Shoreline

Much of the erosion responsible for landforms you might see along the shoreline takes place during storms. Large waves generated by storms release far more energy than normal waves do. This energy is so powerful that it is capable of removing huge chunks of rock. **Figure 4** shows some of the major landscape features that result from wave erosion.

✓ **Reading Check** Why are large waves more capable of removing large chunks of rock from a shoreline than normal waves are?

INTERNET ACTIVITY

For another activity related to this chapter, go to **go.hrw.com** and type in the keyword **HZ5ICEW**.

Headlands are finger-shaped projections that form when cliffs made of hard rock erode more slowly than surrounding rock. On many shorelines, hard rock will form headlands, and the softer rock will form beaches or bays.

Wave-cut terraces form when a sea cliff is worn back, producing a nearly level platform beneath the water at the base of the cliff.

Answer to Reading Check
Large waves are more capable of removing large rocks on a shoreline because they have more energy than normal waves do.

CHAPTER RESOURCES

Technology

📦 **Transparencies**
• Coastal Landforms Created by Wave Erosion: A
• Coastal Landforms Created by Wave Erosion: B

Workbooks

📘 **Science Skills**
• Using Your Senses **GENERAL**

CONNECTION to
Life Science ———— GENERAL

Life on the Shore

PORTFOLIO Beaches and intertidal zones can be challenging places for organisms to live. Beaches offer little protection from predators, and intertidal zones are periodically pounded by waves and exposed to the sun. Most of the organisms that live in these areas have special adaptations for survival. Have students research how different organisms are adapted for living in these environments. Suggest that students use their research to create a diorama that illustrates an intertidal zone.

LS **Verbal/Visual**

BRAIN FOOD

Insuring Ocean Properties
Despite the numerous changes wrought by the ocean, people still live and vacation as close to the water as possible. Inevitably, property is damaged. Government loan subsidies to these property owners can cost taxpayers millions of dollars each year. Encourage students to consider the costs and benefits of erosion prevention. What solutions to the erosion problem would students propose? How would they finance their plans? Allow time for students to share their ideas with the class. **LS** **Logical**

Wave Review Draw a profile of a new shore environment on the board. Have a student volunteer add a diagram of wave energy traveling toward the shore. Ask students to help you identify wave length, wave height, and wave period. Then have students help you illustrate what happens to a wave as it approaches the shore and breaks. Review the movement of water in each part of the diagram. **LS Visual**

Quiz — GENERAL

1. What is a wave period? (It is the time interval between breaking waves.)

2. What determines the way sand moves on a beach? (the direction at which waves strike the shore)

Alternative Assessment — GENERAL

Modeling Coastal Features Have students work independently to make a model of several land features created by waves. Ask students to present their models to the class and explain how the water strikes the shore to create the landforms. Have students brainstorm what organisms, if any, would live on the landforms they modeled. **LS Visual**

England

U.S. Virgin Islands

Hawaii

Figure 5 *Beaches are made of different types of material deposited by waves.*

beach an area of the shoreline made up of material deposited by waves

Wave Deposits

Waves carry a variety of materials, including sand, rock fragments, dead coral, and shells. Often, this material is deposited on a shoreline, where it forms a beach.

Beaches

You would probably recognize a beach if you saw one. However, scientifically speaking, a **beach** is any area of the shoreline made up of material deposited by waves. Some beach material is also deposited by rivers.

Compare the beaches shown in **Figure 5.** Notice that the colors and textures vary. They vary because the type of material found on a beach depends on its source. Light-colored sand is the most common beach material. Much of this sand comes from the mineral quartz. But not all beaches are made of light-colored sand. For example, on many tropical islands, such as the Virgin Islands, beaches are made of fine, white coral material. Some Florida beaches are made of tiny pieces of broken seashells. Black sand beaches in Hawaii are made of eroded volcanic lava. In areas where stormy seas are common, beaches are made of pebbles and boulders.

✓ **Reading Check** Where does beach material come from?

Wave Angle and Sand Movement

The movement of sand along a beach depends on the angle at which the waves strike the shore. Most waves approach the beach at a slight angle and retreat in a direction more perpendicular to the shore. This movement of water is called a longshore current. A *longshore current* is a water current that moves the sand in a zigzag pattern along the beach, as you can see in **Figure 6.**

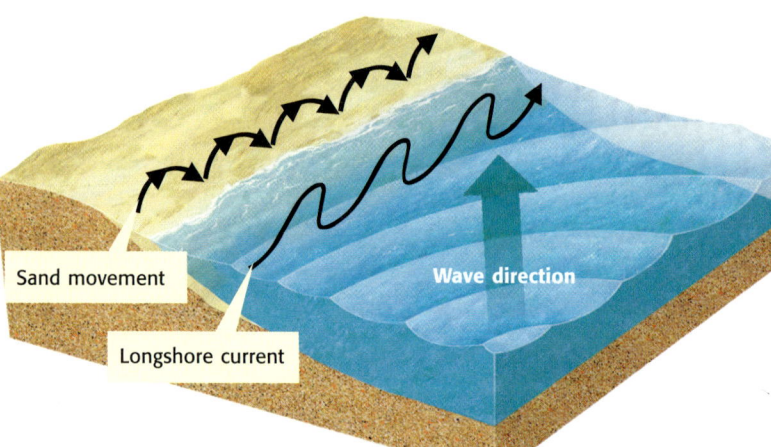

Sand movement

Wave direction

Longshore current

Figure 6 *When waves strike the shoreline at an angle, sand migrates along the beach in a zigzag path.*

Answer to Reading Check

Beach material comes from quartz, coral, broken seashells, lava, pebbles, and boulders.

WEIRD SCIENCE

Sometimes, even when the weather is clear and calm, huge waves, called *rogue waves,* appear unexpectedly. These waves are responsible for damaging or sinking several ships each year. Rogue waves are a poorly understood phenomenon of the high seas. One reason so little is known about rogue waves is that their random nature makes them hard to study.

Offshore Deposits

Waves moving at an angle to the shoreline push water along the shore and create longshore currents. When waves erode material from the shoreline, longshore currents can transport and deposit this material offshore, which creates landforms in open water. A *sandbar* is an underwater or exposed ridge of sand, gravel, or shell material. A *barrier spit* is an exposed sandbar that is connected to the shoreline. Cape Cod, Massachusetts, shown in **Figure 7,** is an example of a barrier spit. A barrier island is a long, narrow island usually made of sand that forms offshore parallel to the shoreline.

Figure 7 *A barrier spit, such as Cape Cod, Massachusetts, occurs when an exposed sandbar is connected to the shoreline.*

SECTION Review

Summary

- As waves break against a shoreline, rock is broken down into sand.
- Six shoreline features created by wave erosion include sea cliffs, sea stacks, sea caves, sea arches, headlands, and wave-cut terraces.
- Beaches are made from material deposited by waves.
- Longshore currents cause sand to move in a zigzag pattern along the shore.

Using Key Terms

Complete each of the following sentences by choosing the correct term from the word bank.

> shoreline   beach

1. A ___ is an area made up of material deposited by waves.

2. An area in which land and a body of water meet is a ___.

Understanding Key Ideas

3. Which of the following is a result of wave deposition?
   a. sea arch
   b. sea cave
   c. barrier spit
   d. headland

4. How do wave deposits affect a shoreline?

5. Describe how sand moves along a beach.

6. What are six shoreline features created by wave erosion?

7. How can the energy of waves traveling through water affect a shoreline?

8. Would a small wave or a large wave have more energy? Explain your answer.

Math Skills

9. Imagine that there is a large boulder on the edge of a shoreline. If the wave period is 15 s long, how many times is the boulder hit in a year?

Critical Thinking

10. **Applying Concepts** Not all beaches are made from light-colored sand. Explain why this statement is true.

11. **Making Inferences** How can severe storms over the ocean affect shoreline erosion and deposition?

12. **Making Predictions** How could a headland change in 250 years? Describe some of the features that may form.

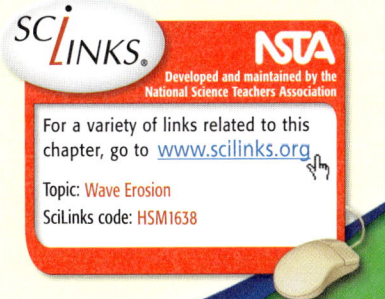

Developed and maintained by the National Science Teachers Association

For a variety of links related to this chapter, go to www.scilinks.org

Topic: Wave Erosion
SciLinks code: HSM1638

CHAPTER RESOURCES

Chapter Resource File

- Section Quiz **GENERAL**
- Section Review **GENERAL**
- Vocabulary and Section Summary **GENERAL**

Focus

Overview

In this section, students learn about the effects of wind erosion. They explore the three major processes of wind erosion: saltation, deflation, and abrasion. Students also learn about the ways that the wind shapes and moves sand dunes.

🔔 Bellringer

Ask students to answer the following question in their **science journal:**

What causes wind? (Students should understand that wind is caused by energy from the sun. The unequal heating of the Earth by the sun causes temperature and pressure differences, which in turn cause air to move.)

Motivate

Discussion —— GENERAL

Wind Erosion Engage students in a discussion about the ways that wind shapes the Earth's surface. Encourage them to compare wind with waves. Ask students to think of examples of the wind's effect on landscapes. (Students should recognize that both wind and waves change the Earth's surface by erosion and deposition. Examples of the wind's effect on landscapes may include dunes and wind-weathered surfaces.) **LS Verbal**

READING WARM-UP

Objectives

● Explain why some areas are more affected by wind erosion than other areas are.

● Describe the process of saltation.

● Identify three landforms that result from wind erosion and deposition.

● Explain how dunes move.

Terms to Learn

saltation   loess
deflation   dune
abrasion

READING STRATEGY

Reading Organizer As you read this section, make a table comparing deflation and abrasion.

Wind Erosion and Deposition

Have you ever been working outside and had a gusty wind blow an important stack of papers all over the place?

Do you remember how fast and far the papers traveled and how long it took to pick them up? Every time you caught up with them, they were on the move again. If this has happened to you, then you have seen how wind erosion works. As an agent of erosion, the wind removes soil, sand, and rock particles and transports them from one place to another.

Certain locations are more vulnerable to wind erosion than others. An area with little plant cover can be severely affected by wind erosion because plant roots anchor sand and soil in place. Deserts and coastlines that are made of fine, loose rock material and have little plant cover are shaped most dramatically by the wind.

The Process of Wind Erosion

Wind moves material in different ways. In areas where strong winds occur, material is moved by saltation. **Saltation** is the skipping and bouncing movement of sand-sized particles in the direction the wind is blowing. As you can see in **Figure 1,** the wind causes the particles to bounce. When moving sand grains knock into one another, some grains bounce up in the air, fall forward, and strike other sand grains. These impacts cause other grains to roll and bounce forward.

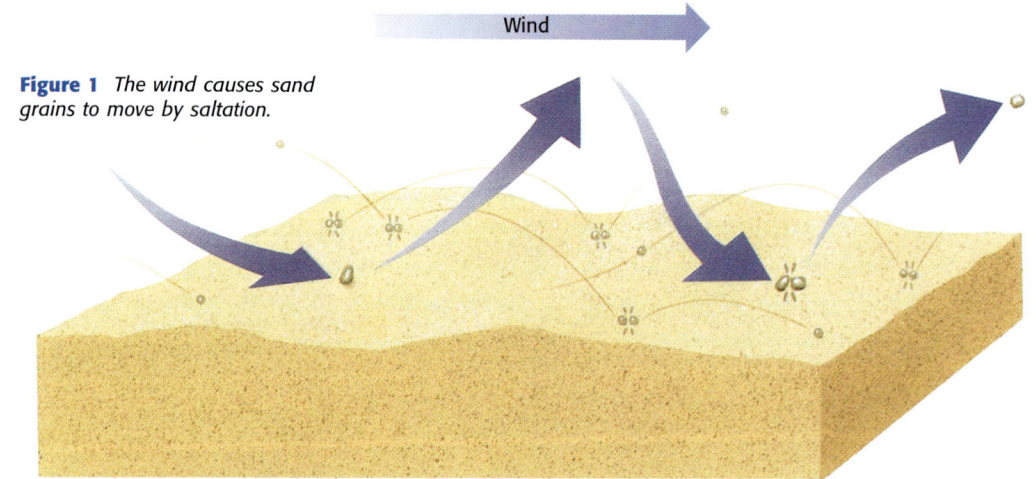

Figure 1 *The wind causes sand grains to move by saltation.*

CHAPTER RESOURCES

Chapter Resource File

📁 • Lesson Plan
   • Directed Reading A **BASIC**
   • Directed Reading B **SPECIAL NEEDS**

Technology

📦 **Transparencies**
   • Bellringer

Figure 2 *Desert pavement, such as that found in the Painted Desert in Arizona, forms when wind removes all the fine materials.*

Deflation

The removal of fine sediment by wind is called **deflation.** During deflation, wind removes the top layer of fine sediment or soil and leaves behind rock fragments that are too heavy to be lifted by the wind. Deflation may cause *desert pavement*, which is a surface consisting of pebbles and small broken rocks. An example of desert pavement is shown in **Figure 2.**

Have you ever blown on a layer of dust while cleaning off a dresser? If you have, you may have noticed that in addition to your face getting dirty, a little scooped-out depression formed in the dust. Similarly, in areas where there is little vegetation, the wind may scoop out depressions in the landscape. These depressions are called *deflation hollows*.

 **Reading Check** Where do deflation hollows form? (*See the Appendix for answers to Reading Checks.*)

Abrasion

The grinding and wearing down of rock surfaces by other rock or sand particles is called **abrasion.** Abrasion commonly happens in areas where there are strong winds, loose sand, and soft rocks. The blowing of millions of sharp sand grains creates a sandblasting effect. This effect helps to erode, smooth, and polish rocks.

saltation the movement of sand or other sediments by short jumps and bounces that is caused by wind or water

deflation a form of wind erosion in which fine, dry soil particles are blown away

abrasion the grinding and wearing away of rock surfaces through the mechanical action of other rock or sand particles

Quick Lab

Making Desert Pavement

1. Spread a mixture of **dust, sand,** and **gravel** on an **outdoor table.**
2. Place an **electric fan** at one end of the table.
3. Put on **safety goggles** and a **filter mask.** Aim the fan across the sediment. Start the fan on its lowest speed. Record your observations.
4. Turn the fan to a medium speed. Record your observations.
5. Finally, turn the fan to a high speed to imitate a desert windstorm. Record your observations.
6. What is the relationship between the wind speed and the size of the sediment that is moved?
7. Does the remaining sediment fit the definition of desert pavement?

CONNECTION to Life Science — GENERAL

Dust in the Wind Each year, equatorial trade winds carry millions of tons of reddish brown dust from the Sahara to Florida. Sahara dust causes hazy skies in Florida and travels as far as South America. The dust provides nutrients for organisms that live in the rain-forest canopies. Traveling over the Pacific Ocean, yellow dust from Mongolia's Gobi Desert reaches Hawaii and fertilizes iron-deficient regions of the Pacific Ocean. In places where the dust settles, plankton populations increase and enrich the food chain. One researcher has tried to link this phenomenon to global climate change. According to this theory, desertification increases during glacial periods, so more sediment is deposited in the oceans. This deposition encourages plankton growth, which removes CO_2 from the atmosphere and further cools the planet.

Section 2 • Wind Erosion and Deposition   **69**

Close

Reteaching — BASIC

Modeling Abrasion
Demonstrate the abrasiveness of sand by briskly rubbing quartz sandpaper on a soft rock specimen, and allow students to observe the changes. Remind students that sandblasting is used in many industrial applications to "erode" hard surfaces. **LS** **Visual**

Quiz — GENERAL

1. Why is sand more likely than silt to move by saltation?
(Sand is heavier than dust and silt, so as sand moves, it tends to bounce along the ground. Silt is light enough to be carried by the wind.)

2. How can the process of deflation form desert pavement?
(Deflation lifts and carries away lighter materials, while the heavier stones remain as desert pavement.)

3. Describe how dunes form.
(When wind encounters an obstacle, the wind slows down and deposits some of the heavier material it is carrying. Gradually, this material collects, becoming a mound and then a dune.)

Alternative Assessment — GENERAL

Concept Mapping Have students use section vocabulary and concepts to construct a concept map that explores the ways wind can shape the Earth's surface. **LS** **Logical**

CONNECTION TO Language Arts

WRITING SKILL **The Dust Bowl** During the 1930s, a severe drought occurred in a section of the Great Plains that became known as the *Dust Bowl*. The wind carried so much dust that some cities left street lights on during the day. Research the Dust Bowl, and describe it in a series of three journal entries written from the perspective of a farmer.

loess very fine sediments deposited by the wind

dune a mound of wind-deposited sand that keeps its shape even though it moves

Wind-Deposited Materials

Much like rivers, the wind also carries sediment. And just as rivers deposit their loads, the wind eventually drops all the material it carries. The amount and the size of particles the wind can carry depend on the wind speed. The faster the wind blows, the more material and the heavier the particles it can carry. As wind speed slows, heavier particles are deposited first.

Loess

Wind can deposit extremely fine material. Thick deposits of this windblown, fine-grained sediment are known as **loess** (LOH ES). Loess feels like the talcum powder a person may use after a shower.

Because wind carries fine-grained material much higher and farther than it carries sand, loess deposits are sometimes found far away from their source. Many loess deposits came from glacial sources during the last Ice Age. In the United States, loess is present in the Midwest, along the eastern edge of the Mississippi Valley, and in eastern Oregon and Washington.

Dunes

When the wind hits an obstacle, such as a plant or a rock, the wind slows down. As it slows, the wind deposits, or drops, the heavier material. The material collects, which creates an additional obstacle. This obstacle causes even more material to be deposited, forming a mound. Eventually, the original obstacle becomes buried. The mounds of wind-deposited sand are called **dunes.** Dunes are common in sandy deserts and along the sandy shores of lakes and oceans. **Figure 3** shows a large dune in a desert area.

Figure 3 *Dunes migrate in the direction of the wind.*

CONNECTION to Life Science — BASIC

Life in the Desert Point out to students that many desert animals have special adaptations to protect themselves from windblown sand. For example, some lizards have transparent eyelids that shield their eyes from blowing sand while still allowing them to see. Encourage students to use library or Internet resources to investigate the organisms that make deserts their home. Have students select one plant or animal and write a brief report describing its habitat, its predators and/or prey, and the adaptations it has developed to live in the desert environment. **English Language Learners** **LS** **Intrapersonal**

The Movement of Dunes

Dunes tend to move in the direction of strong winds. Different wind conditions produce dunes in various shapes and sizes. A dune usually has a gently sloped side and a steeply sloped side, or *slip face,* as shown in **Figure 4.** In most cases, the gently sloped side faces the wind. The wind is constantly transporting material up this side of the dune. As sand moves over the crest, or peak, of the dune, it slides down the slip face, creating a steep slope.

✓ **Reading Check** In what direction do dunes move?

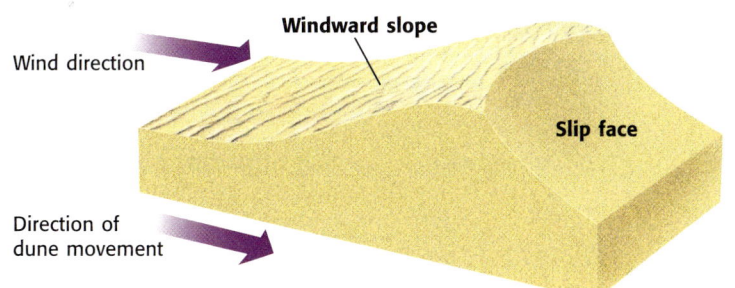

Wind direction

Windward slope

Direction of dune movement

Slip face

Figure 4 *Dunes are formed from material deposited by wind.*

SECTION Review

Summary

- Areas with little plant cover and desert areas covered with fine rock material are more vulnerable than other areas to wind erosion.
- Saltation is the process in which sand-sized particles move in the direction of the wind.
- Three landforms that are created by wind erosion and deposition are desert pavement, deflation hollows, and dunes.
- Dunes move in the direction of the wind.

Using Key Terms

In each of the following sentences, replace the incorrect term with the correct term from the word bank.

> dune   saltation
> deflation   abrasion

1. <u>Deflation hollows</u> are mounds of wind-deposited sand.

2. The removal of fine sediment by wind is called <u>abrasion</u>.

Understanding Key Ideas

3. Which of the following landforms is the result of wind deposition?
   a. deflation hollow
   b. desert pavement
   c. dune
   d. abrasion

4. Describe how material is moved in areas where strong winds blow.

5. Explain the process of abrasion.

Math Skills

6. If a dune moves 40 m per year, how far does it move in 1 day?

Critical Thinking

7. **Identifying Relationships** Explain the relationship between plant cover and wind erosion.

8. **Applying Concepts** If you climbed up the steep side of a sand dune, is it likely that you traveled in the direction the wind was blowing?

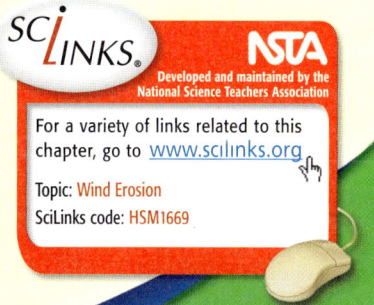

SCLINKS.

NSTA
Developed and maintained by the National Science Teachers Association

For a variety of links related to this chapter, go to www.scilinks.org

Topic: Wind Erosion
SciLinks code: HSM1669

Answer to Reading Check
Dunes move in the direction of strong winds.

CHAPTER RESOURCES

Chapter Resource File

- Section Quiz **GENERAL**
- Section Review **GENERAL**
- Vocabulary and Section Summary **GENERAL**
- Critical Thinking **ADVANCED**
- Datasheet for Quick Lab

Answers to Section Review

1. Dunes are mounds of wind-deposited sand.

2. The removal of fine sediment by wind is called *deflation*.

3. c

4. Sample answer: In areas where strong winds occur, material is moved by saltation. Saltation is the movement of sand-sized particles by bouncing and skipping in the direction that the wind is blowing. The wind lifts sand particles into the air. When the particles land, they hit other particles, which causes them to bounce forward.

5. Sample answer: Abrasion is the grinding and wearing down of rock surfaces by other rock or sand particles. In areas where there are strong winds and loose sand, the blowing of millions of sharp sand grains grinds and wears down rock surfaces.

6. 40 m/year ÷ 365 days/year = 0.11 m/day

7. Sample answer: Vegetation protects areas against the effects of wind erosion. In areas where there is little vegetation, the potential for wind erosion is great.

8. No. The gently sloped side generally faces the wind. The steep side, or slip face, generally faces away from the wind. If you climb up the steep slope, you will generally bc climbing against the wind.

Focus

Overview

This section examines how glaciers form and how they shape the Earth's surface. Students learn to identify different types of glaciers and learn how they move. Students also focus on how glacial erosion and deposition change the Earth's surface.

Bellringer

Tell students that 14,000 years ago, much of North America was covered in a thick layer of ice called a *continental glacier,* which moved as far south as southern Illinois. Humans were living in North America at the time. Have students imagine that they encounter this glacier as an early human, and have them write a paragraph about the experience.

Motivate

Group ACTiViTY — GENERAL

No Glaciers? Divide the class into small groups, and ask the class to imagine an Earth untouched by glaciers. Have the group members work together to make a poster showing such a planet. Encourage them to consider not only Earth's landscape but also the living things inhabiting it. **LS Visual**

READING WARM-UP

Objectives

- Explain the difference between alpine glaciers and continental glaciers.
- Describe two ways in which glaciers move.
- Identify five landscape features formed by alpine glaciers.
- Identify four types of moraines.

Terms to Learn

| glacier | till |
| glacial drift | stratified drift |

READING STRATEGY

Discussion Read this section silently. Write down questions that you have about this section. Discuss your questions in a small group.

glacier a large mass of moving ice

Erosion and Deposition by Ice

Can you imagine an ice cube that is the size of a football stadium? Well, glaciers can be even bigger than that.

A **glacier** is an enormous mass of moving ice. Because glaciers are very heavy and have the ability to move across the Earth's surface, they are capable of eroding, moving, and depositing large amounts of rock materials. And while you will never see a glacier chilling a punch bowl, you might one day visit some of the spectacular landscapes carved by glacial activity!

Glaciers—Rivers of Ice

Glaciers form in areas so cold that snow stays on the ground year-round. In polar regions and at high elevations, snow piles up year after year. Over time, the weight of the snow on top causes the deep-packed snow to become ice crystals. These ice crystals eventually form a giant ice mass. Because glaciers are so massive, the pull of gravity causes them to flow slowly, like "rivers of ice." In this section, you will learn about two main types of glaciers, alpine and continental.

Alpine Glaciers

Alpine glaciers form in mountainous areas. One common type of alpine glacier is a valley glacier. Valley glaciers form in valleys originally created by stream erosion. As these glaciers slowly flow downhill, they widen and straighten the valleys into broad U shapes as shown in **Figure 1.**

✓ Reading Check Where do alpine glaciers form? (*See the Appendix for answers to Reading Checks.*)

Figure 1 *Alpine glaciers start as snowfields in mountainous areas.*

CHAPTER RESOURCES

Chapter Resource File

- Lesson Plan
- Directed Reading A **BASIC**
- Directed Reading B **SPECIAL NEEDS**

Technology

Transparencies
- Bellringer

Answer to Reading Check
Alpine glaciers form in mountainous areas.

Figure 2 *Eleven U.S. states were covered by ice during the last glacial ice period. Because much of the Earth's water was frozen in glaciers, sea levels fell. Blue lines show the coastline at that time.*

Glacial ice
Land

Continental Glaciers

Not all glaciers are true "rivers of ice." In fact, some glaciers spread across entire continents. These glaciers, called *continental glaciers*, are huge, continuous masses of ice. The largest continental glacier in the world covers almost all of Antarctica. This ice sheet is approximately one and a half times the size of the United States. It is so thick—more than 4,000 m in places—that it buries everything but the highest mountain peaks.

Glaciers on the Move

When enough ice builds up on a slope, the ice begins to move downhill. Thick glaciers move faster than thin glaciers, and the steeper the slope is, the faster the glaciers will move. Glaciers move in two ways: by sliding and by flowing. A glacier slides when its weight causes the ice at the bottom of the glacier to melt. As the water from a melting ice cube causes the ice cube to travel across a table, the water from the melting ice causes a glacier to move forward. A glacier also flows slowly as ice crystals within the glacier slip over each other. Think of placing a deck of cards on a table and then tilting the table. The top cards will slide farther than the lower cards. Similarly, the upper part of the glacier flows faster than the base.

Glacier movement is affected by climate. As the Earth cools, glaciers grow. About 10,000 years ago, a continental glacier covered most of North America, as shown in **Figure 2.** In some places, the ice sheet was several kilometers thick!

The *Titanic*

WRITING SKILL An area where an ice sheet is resting on open water is called an *ice shelf*. When pieces of the ice shelf break off, they are called *icebergs.* How far do you think the iceberg that struck the *Titanic* drifted before the two met that fateful night in 1912? Together with a parent, plot on a map of the North Atlantic Ocean the route of the *Titanic* from Southampton, England, to New York. Then, plot a possible route of the drifting iceberg from Greenland to where the ship sank, just south of the Canadian island province of Newfoundland. Describe your findings in your **science journal.**

Is That a Fact!

How do snowflakes become massive blocks of glacial ice? As snow melts and is compacted, the grains become denser. As snow packs to a greater density, the air spaces between ice crystals are pressed out. Eventually, the ice recrystallizes to a stage between flakes and ice called *firn*. Over time, with more pressure from overlying layers of snow, the firn will recrystallize again to become glacial ice.

WEIRD SCIENCE

Glaciers can be very noisy. As they move and stretch, they howl, shriek, pop, groan, and make explosive noises. These sounds are so loud that they have kept high-altitude mountaineers awake at night!

CONNECTION ACTIVITY
Environmental Science — GENERAL

Glaciers in the Water Cycle
Glaciers throughout the world provide fresh water that helps regulate the flow of large rivers and recharge aquifers. Scientists are concerned, however, that a permanent increase in global temperatures would alter this naturally controlled process. If the 13,800,000 km^2 Antarctic ice sheet melted, global sea level could rise 60 m, which would have devastating effects. Coastal towns and cities would be flooded, and some islands would disappear. Have students draw a map of what the coastline of the United States would look like if global sea level rose 60 m. **LS Visual**

MATH PRACTICE

Speed of a Glacier
An alpine glacier is estimated to be moving forward at 5 m per day. Calculate how long the ice will take to reach a road and campground located 0.5 km from the front of the advancing glacier. (Hint: 1 km = 1,000 m)

Landforms Carved by Glaciers

Continental glaciers and alpine glaciers produce landscapes that are very different from one another. Continental glaciers smooth the landscape by scraping and eroding features that existed before the ice appeared. Alpine glaciers carve out rugged features in the mountain rocks through which they flow. **Figure 3** shows the very different landscapes that each type of glacier produces.

Alpine glaciers, such as those in the Rocky Mountains and the Alps, carve out large amounts of rock material and create spectacular landforms. **Figure 4** shows the kinds of landscape features that are sculpted by alpine glaciers.

Figure 3 Landscapes Created by Glaciers

Continental glaciers smooth and flatten the landscape.

Alpine glaciers carved out this rugged landscape.

Is That a Fact!

When metal pipes are drilled through a glacier's layers, they eventually bend in the direction of flow, which demonstrates that glacial layers tend to move at different speeds. One cause of this phenomenon is friction—layers in closest contact with the Earth are often slowed by friction.

CONNECTION to Physical Science — GENERAL

Glacier Movement Point out to students that certain conditions can cause glacial surge. When glaciers surge, or flow rapidly, they may travel 30 m in a day. Explain to students that before and during a glacial surge, meltwater does not drain away from the glacier but builds up beneath the ice, which decreases the friction between the glacier and the land below, permitting the glacier to flow more rapidly.

Figure 4 Landscape Features Carved by Alpine Glaciers

Horns are sharp, pyramid-shaped peaks that form when three or more cirque glaciers erode a mountain.

Cirques (SUHRKS) are bowl-shaped depressions where glacial ice cuts back into the mountain walls.

Arêtes (uh RAYTS) are jagged ridges that form between two or more cirques cutting into the same mountain.

U-shaped valleys form when a glacier erodes a river valley from its original V shape to a U shape.

Hanging valleys are smaller glacial valleys that join the deeper main valley. Many hanging valleys form waterfalls after the ice is gone.

SCIENCE HUMOR

As glaciers travel downward and form U-shaped valleys, outcrops of hard rock may remain on the valley floor. These smooth rocks are called *roches moutonnees*. This French term meaning "sheep rocks" comically describes the rounded outcroppings, which look like sheep grazing in the valley.

CHAPTER RESOURCES

Technology

🗄 **Transparencies**
• Landscape Features Carved by Alpine Glaciers

Workbooks

📘 **Math Skills for Science**
• Using Proportions and Cross-Multiplication GENERAL

READING STRATEGY — BASIC

Preparing Tables Have students work independently to create tables summarizing the characteristics and formation of the following landforms: arêtes, horns, hanging valleys, U-shaped valleys, and cirques. Have students keep the tables to use as study guides.
LS Visual/Verbal

Group ACTIVITY — GENERAL

Making Models Divide the class into pairs, and ask each pair to select a landscape feature created by glaciers. Provide modeling clay for students to make a model of the feature. Have each pair present its model to the class and demonstrate how the feature was formed, using another color of clay to represent the glacier. English Language Learners
LS Visual

CONNECTION ACTIVITY
Language Arts — ADVANCED

Glacial Explorers Students will enjoy reading about the adventures of high-altitude mountaineers and polar explorers. Have students read selections from the accounts of Antarctic explorers such as Robert Falcon Scott, Ernest Shackleton, or Richard Byrd. Students may also enjoy reading about mountaineers such as Reinhold Messner, Sir Edmund Hillary, or Dr. Johan Reinhard. Students should prepare a presentation for the class about the person they studied and should discuss the explorer's description of glaciers. **LS** Verbal

Landscape Features Have students create descriptions in their own words for each landscape feature carved by glaciers. Have students exchange descriptions and quiz each other. **LS Verbal**

1. How are the landscape features formed by continental glaciers different from landscape features formed by alpine glaciers? (Continental glaciers tend to smooth the landscape, whereas alpine glaciers carve out rugged features.)

2. Why is the study of glaciers important? (Answers may vary. Students may note that many of Earth's landforms were created by glacial movement, that glaciers contain much of Earth's fresh water, and that melting glaciers can cause sea levels to rise.)

Alternative Assessment ——— GENERAL

Glacier Handbook Have students create a small glacier handbook that includes illustrations and descriptions of continental and alpine glaciers as well as the types of erosion and deposition they cause. Students' books should include examples of stratified drift, outwash plains, kettles, till, and moraines. **LS Intrapersonal**

glacial drift the rock material carried and deposited by glaciers

till unsorted rock material that is deposited directly by a melting glacier

stratified drift a glacial deposit that has been sorted and layered by the action of streams or meltwater

Types of Glacial Deposits

As a glacier melts, it drops all the material it is carrying. **Glacial drift** is the general term used to describe all material carried and deposited by glaciers. Glacial drift is divided into two main types, *till* and *stratified drift*.

Till Deposits

Unsorted rock material that is deposited directly by the ice when it melts is called **till.** *Unsorted* means that the till is made up of rock material of different sizes—from large boulders to fine sediment. When the glacier melts, the unsorted material is deposited on the surface of the ground.

The most common till deposits are *moraines*. Moraines generally form ridges along the edges of glaciers. Moraines are produced when glaciers carry material to the front of and along the sides of the ice. As the ice melts, the sediment and rock it is carrying are dropped, which forms different types of moraines. The various types of moraines are shown in **Figure 5.**

Figure 5 Types of Moraines

Lateral moraines form along each side of a glacier.

Medial moraines form when valley glaciers with lateral moraines meet.

Ground moraines form from unsorted materials left beneath a glacier.

Terminal moraines form when sediment is dropped at the front of the glacier.

INCLUSION Strategies

• **Behavior Control Issues** • **Learning Disabled**
• **Developmentally Delayed**

Organize students in groups of three to four students to make a mini-glacier. You will need to have access to a freezer. Hand out to each group a plastic cup, gravel, tap water, plastic wrap, strong electrical tape, a paper plate, and a smooth, soft piece of wood. Have each group fill the plastic cup halfway with gravel and then cover the gravel with an inch of water. Carefully tape plastic wrap over the top, and flip the cup over onto the paper plate. Put the cups in the freezer overnight. During the next class, each group should peel off the paper plate and scrape the "glacier" gravel side down across the wood. Ask students what scientists can determine by studying the marks left by the gravel on the wood. Tell students these marks are called *striations* and have students sketch the striation pattern of the glacier in their **science journals.** **LS Visual/Kinesthetic**

Stratified Drift

When a glacier melts, streams form that carry rock material away from the shrinking glacier. A glacial deposit that is sorted into layers based on the size of the rock material is called **stratified drift.** Streams carry sorted material and deposit it in front of the glacier in a broad area called an *outwash plain*. Sometimes, a block of ice is left in the outwash plain when a glacier retreats. As the ice melts, sediment builds up around the block of ice, and a depression called a *kettle* forms. Kettles commonly fill with water to form lakes or ponds, as **Figure 6** shows.

✓ Reading Check Explain the difference between a till deposit and stratified drift.

Figure 6 Kettle lakes form in outwash plains and are common in states such as Minnesota.

SECTION Review

Summary

- Alpine glaciers form in mountainous areas. Continental glaciers spread across entire continents.
- Glaciers can move by sliding or by flowing.
- Alpine glaciers can carve cirques, arêtes, horns, U-shaped valleys, and hanging valleys.
- Two types of glacial drift are till and stratified drift.
- Four types of moraines are lateral, medial, ground, and terminal moraines.

Using Key Terms

Complete each of the following sentences by choosing the correct term from the word bank.

> glacial drift   glacier
> stratified drift   till

1. A glacial deposit that is sorted into layers based on the size of the rock material is called ___.

2. ___ is all of the material carried and deposited by glaciers.

3. Unsorted rock material that is deposited directly by the ice when it melts is ___.

4. A ___ is an enormous mass of moving ice.

Understanding Key Ideas

5. Which of the following is not a type of moraine?
   a. lateral
   b. horn
   c. ground
   d. medial

6. Explain the difference between alpine and continental glaciers.

7. Name five landscape features formed by alpine glaciers.

8. Describe two ways in which glaciers move.

Math Skills

9. A recent study shows that a glacier in Alaska is melting at a rate of 23 ft per year. At what rate is the glacier melting in meters? (Hint: 1 ft = 0.3 m)

Critical Thinking

10. **Analyzing Ideas** Explain why continental glaciers smooth the landscape and alpine glaciers create a rugged landscape.

11. **Applying Concepts** How can a glacier deposit both sorted and unsorted material?

12. **Applying Concepts** Why are glaciers such effective agents of erosion and deposition?

SCLINKS. Developed and maintained by the National Science Teachers Association

For a variety of links related to this chapter, go to www.scilinks.org
Topic: Glaciers
SciLinks code: HSM0675

Focus

Overview

This section introduces gravity as an agent of erosion and deposition. Students learn that mass movements caused by gravity are affected by the material's size, weight, shape, and by the slope on which the material rests. Students then learn about rapid mass movements, such as landslides, mudflows, and volcanic lahars. This section also examines the effect of slow mass movements, such as creep.

🔔 Bellringer

Write the following sentence on the board or overhead projector:

Watch for falling rocks!

Ask students to describe places where a warning sign like this would be necessary. Ask students to consider what factors contribute to make a rock-fall zone.

Motivate

Discussion ——— GENERAL

Erosion by Gravity Ask students to review the three agents of erosion and deposition that they have learned about so far: waves, wind, and glaciers. Then discuss as a class how these types of erosion and deposition compare with erosion and deposition by gravity. **LS Verbal**

The Effect of Gravity on Erosion and Deposition

Did you know that the Appalachian Mountains may have once been almost five times as tall as they are now? Why are they shorter now? Part of the answer lies in the effect that gravity has on all objects on Earth.

Although you can't see it, the force of gravity is also an agent of erosion and deposition. Gravity not only influences the movement of water and ice but also causes rocks and soil to move downslope. **Mass movement** is the movement of any material, such as rock, soil, or snow, downslope. Whether mass movement happens rapidly or slowly, it plays a major role in shaping the Earth's surface.

Angle of Repose

If dry sand is piled up, it will move downhill until the slope becomes stable. The *angle of repose* is the steepest angle, or slope, at which loose material will not slide downslope. This is demonstrated in **Figure 1.** The angle of repose is different for each type of surface material. Characteristics of the surface material, such as its size, weight, shape, and moisture level, determine at what angle the material will move downslope.

READING WARM-UP

Objectives
- Explain the role of gravity as an agent of erosion and deposition.
- Explain how angle of repose is related to mass movement.
- Describe four types of rapid mass movement.
- Describe three factors that affect creep.

Terms to Learn

mass movement   mudflow
rock fall   creep
landslide

READING STRATEGY

Prediction Guide Before reading this section, write the title of each heading in this section. Next, under each heading, write what you think you will learn.

mass movement a movement of a section of land down a slope

Figure 1 *If the slope on which material rests is less than the angle of repose, the material will stay in place. If the slope is greater than the angle of repose, the material will move downslope.*

CHAPTER RESOURCES

Chapter Resource File

- Lesson Plan
- Directed Reading A **BASIC**
- Directed Reading B **SPECIAL NEEDS**

Technology

Transparencies
- Bellringer
- **LINK TO PHYSICAL SCIENCE** Gravitational Force Depends on Mass

Rapid Mass Movement

The most destructive mass movements happen suddenly and rapidly. Rapid mass movement can be very dangerous and can destroy everything in its path.

Rock Falls

While driving along a mountain road, you may have noticed signs along the road that warn of falling rocks. A **rock fall** happens when loose rocks fall down a steep slope. Steep slopes are sometimes created to make room for a road in mountainous areas. Loosened and exposed rocks above the road tend to fall as a result of gravity. The rocks in a rock fall can range in size from small fragments to large boulders.

Landslides

Another type of rapid mass movement is a landslide. A **landslide** is the sudden and rapid movement of a large amount of material downslope. A *slump,* shown in **Figure 2,** is the most common type of landslide. Slumping occurs when a block of material moves downslope over a curved surface. Heavy rains, deforestation, construction on unstable slopes, and earthquakes increase the chances that a landslide will happen. **Figure 3** shows a landslide in India.

✔ **Reading Check** What is a slump? (*See the Appendix for answers to Reading Checks.*)

Figure 2 *A slump is a type of landslide that occurs when a block of land becomes detached and slides downhill.*

rock fall a group of loose rocks that fall down a steep slope

landslide the sudden movement of rock and soil down a slope

Figure 3 *This landslide in Bombay, India, happened after heavy monsoon rains.*

Answer to Reading Check

A slump is the result of a landslide in which a block of material moves downslope over a curved surface.

Flashcards Have students label index cards with the five types of mass movement discussed in this section. On the reverse side of the index cards, have students write details about each type of mass movement. Students can use these flashcards as a study tool. **LS Verbal**

Quiz — GENERAL

1. Name four characteristics of surface material that affect its angle of repose. (size, weight, shape, and moisture level)

2. What is creep? (Creep is the slow movement of surface material downslope.)

3. What are four types of rapid mass movement? (rock falls, landslides, mudflows, and lahars)

Alternative Assessment — GENERAL

Public-Service Announcements Organize the class into groups. Have each group write a public-service announcement designed to educate the public about the dangers of one form of mass movement. Instruct students to focus on the causes and consequences of these phenomena. Have each group present its announcement for the class.
LS Verbal/Kinesthetic

Figure 4 This photo shows one of the many mudflows that have occurred in California during rainy winters.

mudflow the flow of a mass of mud or rock and soil mixed with a large amount of water

Figure 5 This lahar overtook the city of Kyushu in Japan.

Mudflows

A rapid movement of a large mass of mud is a **mudflow.** Mudflows happen when a large amount of water mixes with soil and rock. The water causes the slippery mass of mud to flow rapidly downslope. Mudflows commonly happen in mountainous regions when a long dry season is followed by heavy rains. Deforestation and the removal of ground cover can often result in devastating mudflows. As you can see in **Figure 4,** a mudflow can carry trees, houses, cars, and other objects that lie in its path.

Lahars

Volcanic eruptions or heavy rains on volcanic ash can produce some of the most dangerous mudflows. Mudflows of volcanic origin are called *lahars*. Lahars can travel at speeds greater than 80 km/h and can be as thick as cement. On volcanoes with snowy peaks, an eruption can suddenly melt a great amount of ice. The water from the ice liquefies the soil and volcanic ash to produce a hot mudflow that rushes downslope. **Figure 5** shows the effects of a massive lahar in Japan.

☑ **Reading Check** Explain how a lahar occurs.

Answer to Reading Check
A lahar is caused by the eruption of an ice-covered volcano, which melts ice and causes a hot mudflow.

CONNECTION to Life Science — GENERAL

Mudflows in California In 1980, six successive storms caused devastating mudflows in California. The storms dropped 33 cm of rain, which transformed the soil into mud. Soil on slopes oozed out from under the foundations of houses, which sent the houses tumbling into canyons and valleys, killing 24 people and causing millions of dollars in damage. Many believe that the mudflows were so massive because the area was recently logged.

Slow Mass Movement

Sometimes, you don't even notice mass movement happening. Although rapid mass movements are visible and dramatic, slow mass movements happen a little at a time. However, because slow mass movements occur more frequently, more material is moved collectively over time.

Creep

Even though most slopes appear to be stable, they are actually undergoing slow mass movement, as shown in **Figure 6.** The extremely slow movement of material downslope is called **creep.** Many factors contribute to creep. Water loosens soil and allows it to move freely. In addition, plant roots act as a wedge that forces rocks and soil particles apart. Burrowing animals, such as gophers and groundhogs, also loosen rock and soil particles. In fact, rock and soil on every slope travels slowly downhill.

Figure 6 *Bent tree trunks are evidence that creep is happening.*

creep the slow downhill movement of weathered rock material

SECTION Review

Summary

- Gravity causes rocks and soil to move downslope.
- If the slope on which material rests is greater than the angle of repose, mass movement will occur.
- Four types of rapid mass movement are rock falls, landslides, mudflows, and lahars.
- Water, plant roots, and burrowing animals can cause creep.

Using Key Terms

Complete each of the following sentences by choosing the correct term from the word bank.

creep	mass movement
mudflow	rock fall

1. A ___ occurs when a large amount of water mixes with soil and rock.

2. The extremely slow movement of material downslope is called ___.

Understanding Key Ideas

3. Which of the following is a factor that affects creep?
   a. water
   b. burrowing animals
   c. plant roots
   d. All of the above.

4. How is the angle of repose related to mass movement?

Math Skills

5. If a lahar is traveling at 80 km/h, how long will it take the lahar to travel 20 km?

Critical Thinking

6. **Identifying Relationships** Which types of mass movement are most dangerous to humans? Explain your answer.

7. **Making Inferences** How does deforestation increase the likelihood of mudflows?

SCI LINKS®
Developed and maintained by the National Science Teachers Association

For a variety of links related to this chapter, go to www.scilinks.org

Topic: Mass Movements
SciLinks code: HSM0917

INTERNET ACTIVITY
Essay ——————— GENERAL

For an internet activity related to this chapter, have students go to **go.hrw.com** and type in the keyword **HZ5ICEW.**

CHAPTER RESOURCES

Chapter Resource File
- Section Quiz GENERAL
- Section Review GENERAL
- Vocabulary and Section Summary GENERAL
- SciLinks Activity GENERAL

Gliding Glaciers

Teacher's Notes

Time Required

Two 45-minute class periods plus a 15-minute preparation activity

Lab Ratings

EASY ———————→ HARD

Teacher Prep 🧪🧪
Student Set-Up 🧪🧪
Concept Level 🧪🧪
Clean Up 🧪🧪

MATERIALS

The materials listed on the student page are enough for one student or a pair of students. These materials could also be used in larger groups. To reduce the amount of materials, students could use ice cubes and smaller amounts of clay, sand, and gravel.

Preparation Notes

Students should review the entire section on glaciers in this chapter prior to performing this activity. For the second part of the lab, students might have to refreeze ice blocks overnight or make three more ice blocks. If new ice blocks are made, the sand and gravel can be omitted.

Model-Making Lab

OBJECTIVES

Build a model of a glacier.

Demonstrate the effects of glacial erosion by various materials.

Observe the effect of pressure on the melting rate of a glacier.

MATERIALS

- brick (3)
- clay, modeling (2 lb)
- container, empty large margarine (3)
- freezer
- graduated cylinder, 50 mL
- gravel (1 lb)
- pan, aluminum rectangular (3)
- rolling pin, wood
- ruler, metric
- sand (1 lb)
- stopwatch
- towel, small hand
- water

Gliding Glaciers

A glacier is a large, moving mass of ice. Glaciers are responsible for shaping many of Earth's natural features. Glaciers are set in motion by the pull of gravity and by the gradual melting of the glacier. As a glacier moves, it changes the landscape by eroding the surface over which it passes.

Part A: Getting in the Groove

Procedure

The material that is carried by a glacier erodes Earth's surface by gouging out grooves called *striations*. Different materials have varying effects on the landscape. In this activity, you will create a model glacier with which to demonstrate the effects of glacial erosion by various materials.

1. Fill one margarine container with sand to a depth of 1 cm. Fill another margarine container with gravel to a depth of 1 cm. Leave the third container empty. Fill the containers with water.

2. Put the three containers in a freezer, and leave them there overnight.

3. Retrieve the containers from the freezer, and remove the three ice blocks from the containers.

4. Use a rolling pin to flatten the modeling clay.

5. Hold the ice block from the third container firmly with a towel, and press as you move the ice along the length of the clay. Do this three times. In a notebook, sketch the pattern that the ice block makes in the clay.

Bert Sherwood
Socorro Middle School
El Paso, Texas

CHAPTER RESOURCES

Chapter Resource File

 • Datasheet for Chapter Lab
- Lab Notes and Answers

Technology

 Classroom Videos
- Lab Video

 LabBook

- Dune Movement
- Creating a Kettle

6 Repeat steps 4 and 5 using the ice block that contains sand.

7 Repeat steps 4 and 5 using the ice block that contains gravel.

Analyze the Results

1 **Describing Events** Did any material from the clay become mixed with the material in the ice blocks? Explain.

2 **Describing Events** Was any material from the ice blocks deposited on the clay surface? Explain.

3 **Examining Data** What glacial features are represented in your clay model?

Draw Conclusions

4 **Evaluating Data** Compare the patterns formed by the three model glaciers. Do the patterns look like features carved by alpine glaciers or by continental glaciers? Explain.

Part B: Melting Away

Procedure

As the layers of ice build up and a glacier gets larger, a glacier will eventually begin to melt. The water from the melted ice allows a glacier to move forward. In this activity, you'll explore the effect of pressure on the melting rate of a glacier.

1 If possible, make three identical ice blocks without any sand or gravel in them. If that is not possible, use the ice blocks from Part A. Place one ice block upside down in each pan.

2 Place one brick on top of one of the ice blocks. Place two bricks on top of another ice block. Do not put any bricks on the third ice block.

3 After 15 min, remove the bricks from the ice blocks.

4 Using the graduated cylinder, measure the amount of water that has melted from each ice block.

5 Observe and record your findings.

Analyze the Results

1 **Analyzing Data** Which ice block produced the most water?

2 **Explaining Events** What did the bricks represent?

3 **Analyzing Results** What part of the ice blocks melted first? Explain.

Draw Conclusions

4 **Interpreting Information** How could you relate this investigation to the melting rate of glaciers? Explain.

Applying Your Data

Replace the clay with different materials, such as soft wood or sand. How does each ice block affect the different surface materials? What types of surfaces do the different materials represent?

CHAPTER RESOURCES

Workbooks

 Whiz-Bang Demonstrations
- Between a Rock and a Hard Place **BASIC**
- Rising Mountains **GENERAL**

 Long-Term Projects & Research Ideas
- Deep in the Mud **ADVANCED**

Lab Notes

This part of the lab models how the weight of glacial ice causes the ice at the bottom of the glacier to melt. This is one way that glaciers move. You may wish to explain this concept by discussing how ice skates work. Ice skates glide smoothly because they distribute a skater's weight on two thin blades. The weight of a skater applied to such a small surface area causes the ice beneath the blades of the skate to melt and quickly re-freeze. Like glaciers, ice skaters glide on a thin layer of water.

Part A: Analyze the Results

1. Answers may vary. Small amounts of the surface material may become mixed with the ice.

2. Answers may vary. Small amounts of the material in the ice may be deposited on the clay surface.

3. Answers may vary. Answers may include moraines, striations, and outwash plains.

Part A: Draw Conclusions

4. Answers may vary. Alpine glaciers leave rugged features behind as they flow. Continental glaciers smooth the landscape.

Part B: Analyze the Results

1. The ice block with two bricks on it produced the most water.

2. The bricks represented layers of ice.

3. The bottom of the ice block melted first because of the weight of the bricks on top of the ice block.

Part B: Draw Conclusions

4. Students should conclude that glaciers that are heavier melt faster. This, in turn, causes glaciers to move faster.

Assignment Guide

SECTION	QUESTIONS
1	1, 2, 7–9, 15, 24–27
2	3, 10, 16, 18, 20, 21
3	4, 5, 11, 12, 19, 22, 23
4	6, 13, 14, 17

ANSWERS

Using Key Terms

1. Sample answer: The shoreline is the area where land and a body of water meet. A longshore current is a movement of water close to the shoreline that moves sand in a zigzag pattern.

2. Sample answer: Beaches are areas of the shoreline made up of material deposited by waves. Dunes are deposits of windblown sand that can be found on a beach.

3. Sample answer: Deflation is the lifting and removal of material by the wind. Saltation is the movement of sand by a skipping and bouncing action in the direction the wind is blowing.

4. Sample answer: A continental glacier is a large, continuous mass of ice that can spread across an entire continent. Alpine glaciers form in mountainous areas, and they are much smaller than continental glaciers. Continental glaciers tend to smooth out the landscape, while alpine glaciers tend to carve rugged features in mountains.

USING KEY TERMS

For each pair of terms, explain how the meanings of the terms differ.

1. *shoreline* and *longshore current*

2. *beaches* and *dunes*

3. *deflation* and *saltation*

4. *continental glacier* and *alpine glacier*

5. *stratified drift* and *till*

6. *mudflow* and *creep*

UNDERSTANDING KEY IDEAS

Multiple Choice

7. *Surf* refers to
   a. large storm waves in the open ocean.
   b. giant waves produced by hurricanes.
   c. breaking waves near the shoreline.
   d. small waves on a calm sea.

8. When waves cut completely through a headland, a ___ is formed.
   a. sea cave
   b. sea arch
   c. wave-cut terrace
   d. sandbar

9. A narrow strip of sand that is formed by wave deposition and is connected to the shore is called a
   a. barrier spit.
   b. sandbar.
   c. wave-cut terrace.
   d. headland.

10. A wind-eroded depression is called a
    a. deflation hollow.
    b. desert pavement.
    c. dune.
    d. dust bowl.

11. What term describes all types of glacial deposits?
    a. glacial drift
    b. dune
    c. till
    d. outwash

12. Which of the following is NOT a landform created by an alpine glacier?
    a. cirque
    b. deflation hollow
    c. horn
    d. arête

13. What is the term for a mass movement that is of volcanic origin?
    a. lahar
    b. slump
    c. creep
    d. rock fall

14. Which of the following is a slow mass movement?
    a. mudflow
    b. landslide
    c. creep
    d. rock fall

Short Answer

15. Why do waves break when they near the shore?

16. Why are some areas more affected by wind erosion than other areas are?

5. Sample answer: Stratified drift is sorted glacial drift. Till is unsorted glacial drift.

6. Sample answer: A mudflow is a rapid mass movement. Creep is a slow mass movement.

Understanding Key Ideas

7. c	**11.** a
8. b	**12.** b
9. a	**13.** a
10. a	**14.** c

15. When waves reach shallow water, the lower part of the wave is crowded by the ocean floor. The wave becomes taller and eventually grows so tall that it cannot support itself. When the wave reaches this point, it curls and breaks.

16. Areas with little vegetation, deserts, and coastlines are more affected by wind erosion than other areas because there are few plant roots to anchor the sand and soil in place.

17 What kind of mass movement happens continuously, day after day?

18 In what direction do sand dunes move?

19 Describe the different types of glacial moraines.

CRITICAL THINKING

20 Concept Mapping Use the following terms to create a concept map: *deflation, strong winds, saltation, dune,* and *desert pavement.*

21 Making Inferences How do humans increase the likelihood that wind erosion will occur?

22 Identifying Relationships If the large ice sheet covering Antarctica were to melt completely, what type of landscape would you expect Antarctica to have?

23 Applying Concepts You are a geologist who is studying rock to determine the direction of flow of an ancient glacier. What clues might help you determine the glacier's direction of flow?

24 Applying Concepts You are interested in purchasing a home that overlooks the ocean. The home that you want to buy sits atop a steep sea cliff. Given what you have learned about shoreline erosion, what factors would you take into consideration when deciding whether to buy the home?

INTERPRETING GRAPHICS

The graph below illustrates coastal erosion and deposition at an imaginary beach over a period of 8 years. Use the graph below to answer the questions that follow.

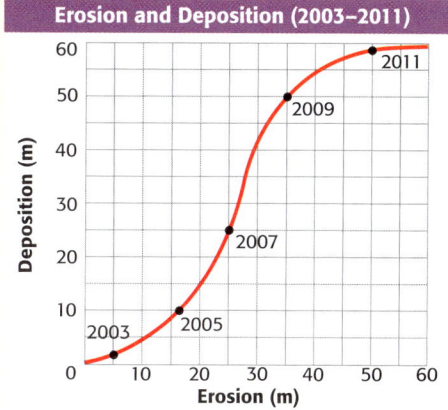

Erosion and Deposition (2003–2011)

25 What is happening to the beach over time?

26 In what year does the amount of erosion equal the amount of deposition?

27 Based on the erosion and deposition data for 2005, what might happen to the beach in the years that follow 2005?

CHAPTER RESOURCES

Chapter Resource File

- Chapter Review **GENERAL**
- Chapter Test A **GENERAL**
- Chapter Test B **ADVANCED**
- Chapter Test C **SPECIAL NEEDS**
- Vocabulary Activity **GENERAL**

Workbooks

Study Guide
- Assessment resources are also available in Spanish.

  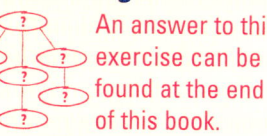

Standardized Test Preparation

Teacher's Note

To provide practice under more realistic testing conditions, give students 20 minutes to answer all of the questions in this Standardized Test Preparation.

MISCONCEPTION ALERT

Answers to the standardized test preparation can help you identify student misconceptions and misunderstandings.

READING

Passage 1

1. B
2. H
3. A
4. I

✚ **TEST DOCTOR**

Question 3: Some students may think that taller waves are formed in the deeper water because the depth allows for larger waves. However, as the passage explains, waves become crowded as they reach the shore and therefore get taller.

Question 4: Some students may misinterpret short waves that are far apart as the waves that break on the shore. However, the passage implies that in the deep ocean water, waves have more room and are therefore shorter and farther apart than waves that are moving closer to the shore.

READING

Read each of the passages below. Then, answer the questions that follow each passage.

Passage 1 When you drop a pebble into a pond, is there just one ripple? Of course not. Waves, like ripples, don't move alone. Waves travel in groups called wave <u>trains</u>. As wave trains move away from their sources, they travel through the ocean water <u>uninterrupted</u>. But when waves reach shallow water, they change form because the ocean floor crowds the lower part of the wave. As a result, the waves get closer together and taller.

1. In this passage, what does the word *uninterrupted* mean?
   A not continuous
   B not broken
   C broken again
   D not interpreted

2. In this passage, what does the word *train* mean?
   F to teach someone a skill
   G the part of a gown that trails behind the person who is wearing the gown
   H a series of moving things
   I a series of railroad cars

3. According to the passage, what is the cause of taller waves?
   A shallow water
   B deep ocean water
   C rippling
   D wave trains

4. If certain waves are short and far apart, which of the following can be concluded?
   F The waves are approaching the shore.
   G The waves are moving toward their source.
   H The waves were interrupted.
   I The waves are in deep ocean water.

Passage 2 Winter storms create powerful waves that crash into cliffs and break off pieces of rock that fall into the ocean. On February 8, 1998, unusually large waves crashed against the cliffs along Broad Beach Road in Malibu, California. Eventually, the ocean-eroded cliffs <u>buckled</u>, which caused a landslide. One house collapsed into the ocean, and two more houses dangled on the edge of the cliff's newly eroded face. Powerful waves, buckled cliffs, and landslides are part of the ongoing natural process of coastal erosion that is taking place along the California shoreline and along similar shorelines throughout the world.

1. In this passage, what does *buckled* mean?
   A tightened
   B collapsed
   C formed
   D heated up

2. Which of the following describes how this coastal area was damaged?
   F The area was damaged by collapsing houses.
   G The area was damaged an earthquake.
   H The area was damaged by ocean currents.
   I The area was damaged by unusually large waves produced by a winter storm.

3. Which of the following can be concluded from this passage?
   A This area may have landslides in the future.
   B This area is safe from future landslides.
   C This type of landslide is common only to the California coastline.
   D Erosion in this area happens very rarely.

Passage 2

1. B
2. I
3. A

✚ **TEST DOCTOR**

Question 3: Some students may think because this area was severely eroded from the storm and landslide discussed in the passage, the area may be safe from future landslides. In fact, the passage infers that the area is susceptible to future landslides because of its location along the shoreline.

Use each figure below to answer the questions that follow each figure.

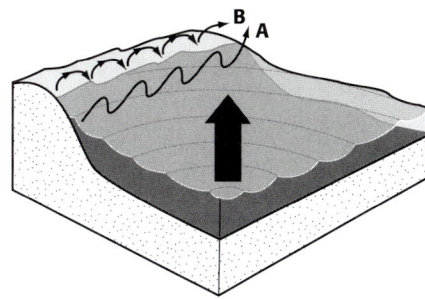

1. In the illustration, what does A label?
   A wave direction
   B wave amplitude
   C wavelength
   D a longshore current

2. In the illustration, what does B label?
   F wave direction
   G wave period
   H the movement of sand
   I a longshore current

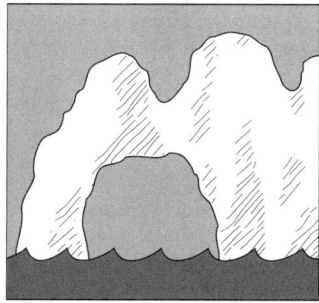

3. What process created the landform in the illustration above?
   A erosion by waves
   B saltation
   C abrasion
   D deposition by waves

Read each question below, and choose the best answer.

1. Wind erosion caused a deflation hollow that was circular in shape. The hollow is 100 m wide. What is the circumference of this deflation hollow?
   A 31.4 m
   B 62.8 m
   C 314 m
   D 628 m

2. A homeowner needs to buy and plant 28 trees to prevent wind erosion. Each tree costs $29.99. What is a reasonable estimate for the total cost of these trees before tax?
   F a little more than $200
   G a little less than $600
   H a little less than $900
   I a little more than $1,000

Use the equation below to answer the questions that follow.

$$\frac{number\ of\ waves}{per\ minute} = \frac{60\ s}{wave\ period\ (s)}$$

3. If the wave period is 15 s, how many waves occur in 1 min?
   A 4
   B 60
   C 75
   D 240

4. If the wave period is 30 s, how many waves occur in 1 min?
   F 1
   G 2
   H 3
   I 5

5. If 480 waves broke in 40 min, what is the wave period?
   A 5 s
   B 12 s
   C 15 s
   D 20 s

Standardized Test Preparation

INTERPRETING GRAPHICS
1. D
2. H
3. A

+ TEST DOCTOR

Question 3: Saltation refers to the movement of sand along the shoreline. Abrasion refers to a form of wind erosion that smooths and polishes rock. This diagram shows a sea arch, which is formed by wave erosion rather than deposition, so answer A is correct.

MATH
1. C
2. I
3. A
4. G
5. A

+ TEST DOCTOR

Question 3: Answer B is the number of waves that would occur in 15 minutes. Answer C is the number of waves that would occur in 19 minutes. Answer D is the number of waves that would occur in 1 hour. So, answer A is the correct answer.

CHAPTER RESOURCES

Chapter Resource File
 • Standardized Test Preparation GENERAL

State Resources
 For specific resources for your state, visit **go.hrw.com** and type in the keyword **HSMSTR**.

Science in Action

Weird Science

Background

Friction calculations predict that the ratio of a landslide's vertical drop to the horizontal distance it travels should be about 0.6 to 1. For some rare, giant landslides of rock, however, this ratio can be as low as 0.1 to 1. These long-runout landslides are also known as "sturzstroms," which is German for "fall streams."

Scientific Discoveries

Background

The recovery team looked at the ice near the surface and saw about 30 cm of ice between each year's summer melt. The team also reasoned that global warming may have increased melting and slowed the rate of deposition. With this information, they expected the planes to be buried under 12 m of ice. So why were they so far off?

Studies over the last few decades show a slight rise in global temperatures, but the area of the crash site has actually been cooling. This means less melting each summer, so more snow is added to the growing layers of ice. Also, rising global temperatures may be causing more snowfall in some parts of Greenland.

Weird Science

Long-Runout Landslides

At 4:10 A.M. on April 29, 1903, the town of Frank, Canada, was changed forever when disaster struck without warning. An enormous chunk of limestone fell suddenly from the top of nearby Turtle Mountain. In less than two minutes, the huge mass of rock buried most of the town! Landslides such as the Frank landslide are now known as *long-runout landslides*. Most landslides travel a horizontal distance that is less than twice the vertical distance that they have fallen. But long-runout landslides carry enormous amounts of rock and thus can travel many times farther than they fall. The physics of long-runout landslides are still a mystery to scientists.

Scientific Discoveries

The Lost Squadron

During World War II, an American squadron of eight planes crash-landed on the ice of Greenland. The crew was rescued, but the planes were lost. After the war, several people tried to find the "Lost Squadron." Finally, in 1988, a team of adventurers found the planes by using radar. The planes were buried by 40 years of snowfall and had become part of the Greenland ice sheet! When the planes were found, they were buried under 80 m of glacial ice. Incredibly, the team tunneled down through the ice and recovered a plane. The plane is now named Glacier Girl, and it still flies today!

Language Arts ACTIVITY

WRITING SKILL The crew of the Lost Squadron had to wait 10 days to be rescued by dog sled. Imagine that you were part of the crew—what would you have done to survive? Write a short story describing your adventure on the ice sheet of Greenland.

Math ACTIVITY

The Frank landslide traveled 4 km in 100 s. Calculate this speed in meters per second.

Answer to Math Activity

4 km = 4,000 m

4,000 m ÷ 100 s = 40 m/s

Answer to Language Arts Activity

Answers may vary. Students' stories should describe harsh weather conditions, similar to the conditions that the Lost Squadron crew encountered, and the stories could provide a description of the survival techniques the students depended upon until their crew was rescued.

8. Viewing the potato from above, use the transparency marker to trace the outline of the potato where it rests on the bottom of the container. The floor of the container corresponds to the summer water level in the lake.

9. Label this contour "0 m." (For this activity, assume that the water level in the lake during the summer is the same as sea level.)

10. Pour water into the container until it reaches the line labeled "1 cm."

11. Again, place the lid on the container, and seal it. Part of the potato will be sticking out above the water. Viewing the potato from above, trace the part of the potato that touches the top of the water.

12. Label the elevation of the contour line you drew in step 11. According to the scale, the elevation is 10 m.

13. Remove the lid. Carefully pour water into the container until it reaches the line labeled "2 cm."

14. Place the lid on the container, and seal it. Viewing the potato from above, trace the part of the potato that touches the top of the water at this level.

15. Use the scale to calculate the elevation of this line. Label the elevation on your drawing.

16. Repeat steps 13–15, adding 1 cm to the depth of the water each time. Stop when the potato is completely covered.

17. Remove the lid, and set it on a tabletop. Place tracing paper on top of the lid. Trace the contours from the lid onto the paper. Label the elevation of each contour line. Congratulations! You have just made a topographic map!

Analyze the Results

1. What is the contour interval of this topographic map?

2. By looking at the contour lines, how can you tell which parts of the potato are steeper?

3. What is the elevation of the highest point on your map?

Draw Conclusions

4. Do all topographic maps have a 0 m elevation contour line as a starting point? How would this affect a topographic map of Sometimes Island? Explain your answer.

5. Would this method of measuring elevation be an effective way to make a topographic map of an actual area on Earth's surface? Why or why not?

> **Applying Your Data**
>
> Place all of the potatoes on a table or desk at the front of the room. Your teacher will mix up the potatoes as you trade topographic maps with another group. By reading the topographic map you just received, can you pick out the matching potato?

Analyze the Results

1. The contour interval of the topographic map is 10 m.

2. Steeper parts of the potato will have contour lines that are closer together.

3. Answers may vary. Elevation is indicated by numbers on the contour lines.

Draw Conclusions

4. Sample answer: no; Topographic maps do not necessarily start with a 0 m elevation contour line. It is possible to make a topographic map of Sometimes Island showing its contours, but it is impossible to know the island's elevation above sea level.

5. Sample answer: no; Flooding an island is not an effective way of mapping it.

Michael E. Kral
West Hardin Middle School
Cecilia, Kentucky

CHAPTER RESOURCES
Chapter Resource File
• Datasheet for LabBook
• Lab Notes and Answers

Great Ice Escape

Teacher's Notes

Time Required
Two 45-minute class periods

Lab Ratings

EASY ————————→ HARD

Teacher Prep 🧪🧪🧪
Student Set-Up 🧪🧪
Concept Level 🧪
Clean Up 🧪🧪

MATERIALS
The materials listed on the student page are enough for 1 to 3 students.

Safety Caution
Remind students to review all safety cautions and icons before beginning this lab activity. Warn them that when a plastic jar cracks, the pieces could be very sharp.

Preparation Notes
You should perform this lab ahead of time in order to make certain that your plastic jars will break.

Test the Hypothesis
8. Answers may vary.
9. Sample answer: Ice is protruding from the top of the unsealed jar. Both sealed jars are cracked. The unwrapped jar cracked more severely.
10. Answers may vary, students should observe that the height of the water has increased.

Great Ice Escape

Did you know that ice acts as a natural wrecking ball? Even rocks don't stand a chance against the power of ice. When water trapped in rock freezes, a process called *ice wedging* occurs. The water volume increases, and the rock cracks to "get out of the way." This expansion can fragment a rock into several pieces. In this exercise, you will see how this natural wrecker works, and you will try to stop the great ice escape.

Ask a Question
1. If a plastic jar is filled with water, is there a way to prevent the jar from breaking when the water freezes?

Form a Hypothesis
2. Write a hypothesis that is a possible answer to the question above. Explain your reasoning.

Test the Hypothesis
3. Fill three identical jars to overflowing with water, and close two of them securely.
4. Measure the height of the water in the unsealed container. Record the height.
5. Tightly wrap one of the closed jars with tape, string, or other items to reinforce the jar. These items must be removable.
6. Place all three jars in resealable sandwich bags, and leave them in the freezer overnight. (Make sure the open jar does not spill.)
7. Remove the jars from the freezer, and carefully remove the wrapping from the reinforced jar.
8. Did your reinforced jar crack? Why or why not?
9. What does each jar look like? Record your observations.
10. Record the height of the ice in the unsealed jar. How does the new height compare with the height you measured in step 4?

Analyze the Results
1. Do you think it is possible to stop the ice from breaking the sealed jars? Why or why not?
2. How could ice wedging affect soil formation?

MATERIALS
- bags, sandwich resealable (3)
- freezer
- jars, hard plastic with screw-on lids, such as spice containers (3)
- ruler, metric
- tape, strings, rubber bands, and other items to bind or reinforce the jars
- water

SAFETY

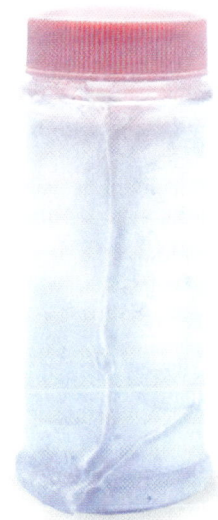

Analyze the Results
1. Sample answer: no; The expanding ice cannot be confined. If it can shatter a rock, it can break plastic.
2. Sample answer: Ice wedging breaks up large rocks into smaller pieces that are further weathered chemically and physically. The processes of weathering create soil.

CHAPTER RESOURCES

 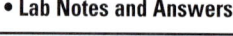

Chapter Resource File
- Datasheet for LabBook
- Lab Notes and Answers

David M. Sparks
Redwater Junior High School
Redwater, Texas

Model-Making Lab

Dune Movement

Wind moves the sand by a process called *saltation*. The sand skips and bounces along the ground in the same direction as the wind is blowing. As sand is blown across a beach, the dunes change. In this activity, you will investigate the effect wind has on a model sand dune.

Procedure

1. Use the marker to draw and label vertical lines 5 cm apart along one side of the box.

2. Fill the box about halfway with sand. Brush the sand into a dune shape about 10 cm from the end of the box.

3. Use the lines you drew along the edge of the box to measure the location of the dune's peak to the nearest centimeter.

4. Slide the box into the paper bag until only about half the box is exposed, as shown below.

5. Put on your safety goggles and filter mask. Hold the hair dryer so that it is level with the peak of the dune and about 10–20 cm from the open end of the box.

6. Turn on the hair dryer at the lowest speed, and direct the air toward the model sand dune for 1 min.

7. Record the new location of the model dune.

8. Repeat steps 5 and 6 three times. After each trial, measure and record the location of the dune's peak.

Analyze the Results

1. How far did the dune move during each trial?

2. How far did the dune move overall?

Draw Conclusions

3. How might the dune's movement be affected if you were to turn the hair dryer to the highest speed?

Applying Your Data

Flatten the sand. Place a barrier, such as a rock, in the sand. Position the hair dryer level with the top of the sand's surface. How does the rock affect the dune's movement?

MATERIALS

- bag, paper, large enough to hold half the box
- box, cardboard, shallow
- hair dryer
- marker
- mask, filter
- ruler, metric
- sand, fine

SAFETY

Model-Making Lab

Dune Movement

Teacher's Notes

Time Required
30 minutes

Lab Ratings

EASY ————————— HARD

Teacher Prep ⚗
Student Set-Up ⚗⚗
Concept Level ⚗
Clean Up ⚗⚗

MATERIALS
The materials listed on the student page are enough for 2 students.

Safety Caution
Remind students to review all safety cautions and icons before beginning this lab activity.

Preparation Notes
You might want to have students do this activity outside, in an area where an electrical outlet is available.

Analyze the Results

1. Answers may vary. A typical answer would be about 0.5 to 1.0 cm.

2. Answers may vary. A typical answer would be about 5 to 10 cm.

Draw Conclusions

3. Answers may vary. The sand could be blown until it hits the bag, or the dune could move farther. Students may also have predicted that more sand would be blown out of the box.

Applying Your Data
The dune forms on the downwind side of the barrier. The rock slows the migration of the dune.

CHAPTER RESOURCES

Chapter Resource File
- Datasheet for LabBook
- Lab Notes and Answers

CLASSROOM TESTED & APPROVED

Larry Tackett
Andrew Jackson Middle School
Cross Lanes, West Virginia

Creating a Kettle

Teacher's Notes

Time Required

One 45-minute class period plus 30 minutes during a second day

Lab Ratings

EASY ———————— HARD

Teacher Prep 🧪
Student Set-Up 🧪
Concept Level 🧪
Clean Up 🧪

MATERIALS

The materials listed on the student page are enough for a group of 4 to 5 students.

Analyze the Results

1. Sample answer: The model is similar. The simulated kettle is the size of the ice cube. A real kettle hole is the size of the block of ice that breaks off a glacier. Its shape is determined by the shape of the ice.

2. Sample answer:
Similarities: The ice cube in the lab melted slowly to form a depression. Similarly, blocks of ice left behind by glaciers melt slowly to form kettles.

Creating a Kettle

As glaciers recede, they leave huge amounts of rock material behind. Sometimes receding glaciers form moraines by depositing some of the rock material in ridges. At other times, glaciers leave chunks of ice that form depressions called *kettles*. As the ice melts, these depressions may form ponds or lakes. In this activity, you will discover how kettles are formed by creating your own.

MATERIALS

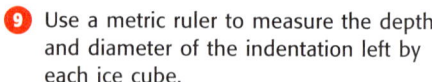

- ice, cubes of various sizes (4–5)
- ruler, metric
- sand
- tub, small

Ask a Question

1 How are kettles formed?

Form a Hypothesis

2 Write a hypothesis that could answer the question above.

Test the Hypothesis

3 Fill the tub three-quarters full with sand.

4 Describe the size and shape of each ice cube.

5 Push the ice cubes to various depths in the sand.

6 Put the tub where it won't be disturbed overnight.

7 Closely observe the sand around the area where you left each ice cube.

8 What happened to the ice cubes?

9 Use a metric ruler to measure the depth and diameter of the indentation left by each ice cube.

Analyze the Results

1 How does this model relate to the size and shape of a natural kettle?

2 In what ways are your model kettles similar to real ones? How are they different?

Draw Conclusions

3 Based on your model, what can you conclude about the formation of kettles by receding glaciers?

Differences: The materials and debris surrounding the model hole are different from those surrounding a real kettle. In addition, the shape of a real kettle would not be as uniform as the shape of the model hole.

Draw Conclusions

3. Answers may vary. Accept all reasonable responses. Kettles form from the slow melting of ice left behind when a glacier recedes.

CHAPTER RESOURCES

Chapter Resource File

- Datasheet for LabBook
- Lab Notes and Answers

CLASSROOM TESTED & APPROVED

Janel Guse
West Central Middle School
Hartford, South Dakota

✓ Reading Check Answers

Chapter 1   Maps as Models of the Earth
Section 1
Page 5: A reference point is a fixed place on the Earth's surface from which direction and location can be described.

Page 6: True north is the direction to the geographic North Pole.

Page 8: lines of longitude

Section 2
Page 10: Distortions are inaccuracies produced when information is transferred from a curved surface to a flat surface.

Page 13: Azimuthal and conic projections are similar because they are both ways to represent the curved surface of the Earth on a flat map. Azimuthal projections show the surface of a globe transferred to a flat plane, whereas conic projections show the surface of a globe transferred to a cone.

Page 14: Every map should have a title, a compass rose, a scale, the date, and a legend.

Page 16: A GIS stores information in layers.

Section 3
Page 19: An index contour is a darker contour line that is usually every fifth line. Index contours make it easier to read a map.

Chapter 2   Weathering and Soil Formation
Section 1
Page 33: Wind, water, and gravity can cause abrasion.

Page 34: Answers may vary. Sample answer: ants, worms, mice, coyotes, and rabbits.

Page 37: Oxidation occurs when oxygen combines with an element to form an oxide.

Section 2
Page 39: As the surface area increases, the rate of weathering also increases.

Page 40: Warm, humid climates have higher rates of weathering because oxidation happens faster when temperatures are higher and when water is present.

Page 41: Mountains weather faster because they are exposed to more wind, rain, and ice, which are agents of weathering.

Section 3
Page 42: Soil is formed from parent rock, organic material, water, and air.

Page 45: Heavy rains leach precious nutrients into deeper layers of soil, resulting in a very thin layer of topsoil.

Page 46: Temperate climates have the most productive soil.

Section 4
Page 48: Soil provides nutrients to plants, houses for animals, and stores water.

Page 51: They restore important nutrients to the soil and provide cover to prevent erosion.

Chapter 3   Agents of Erosion and Deposition
Section 1
Page 63: The amount of energy released from breaking waves causes rock to break down, eventually forming sand.

Page 65: Large waves are more capable of moving large rocks on a shoreline because they have more energy than normal waves do.

Page 66: Beach material is material deposited by waves.

Section 2
Page 69: Deflation hollows form in areas where there is little vegetation.

Page 71: Dunes move in the direction of strong winds.

Section 3
Page 72: Alpine glaciers form in mountainous areas.

Page 77: A till deposit is made up of unsorted material, while stratified drift is made up of sorted material.

Section 4
Page 79: A slump is the result of a landslide in which a block of material moves downslope over a curved surface.

Page 80: A lahar is caused by the eruption of an ice-covered volcano, which melts ice and causes a hot mudflow.

Study Skills

FoldNote Instructions

Have you ever tried to study for a test or quiz but didn't know where to start? Or have you read a chapter and found that you can remember only a few ideas? Well, FoldNotes are a fun and exciting way to help you learn and remember the ideas you encounter as you learn science!

FoldNotes are tools that you can use to organize concepts. By focusing on a few main concepts, FoldNotes help you learn and remember how the concepts fit together. They can help you see the "big picture." Below you will find instructions for building 10 different FoldNotes.

Pyramid

1. Place a sheet of paper in front of you. Fold the lower left-hand corner of the paper diagonally to the opposite edge of the paper.

2. Cut off the tab of paper created by the fold (at the top).

3. Open the paper so that it is a square. Fold the lower right-hand corner of the paper diagonally to the opposite corner to form a triangle.

4. Open the paper. The creases of the two folds will have created an X.

5. Using scissors, cut along one of the creases. Start from any corner, and stop at the center point to create two flaps. Use tape or glue to attach one of the flaps on top of the other flap.

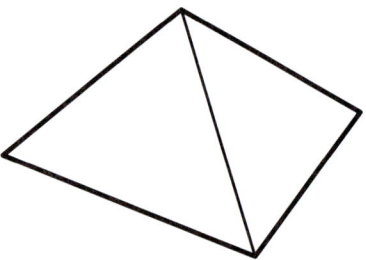

Double Door

1. Fold a sheet of paper in half from the top to the bottom. Then, unfold the paper.

2. Fold the top and bottom edges of the paper to the crease.

Booklet

1. Fold a sheet of paper in half from left to right. Then, unfold the paper.

2. Fold the sheet of paper in half again from the top to the bottom. Then, unfold the paper.

3. Refold the sheet of paper in half from left to right.

4. Fold the top and bottom edges to the center crease.

5. Completely unfold the paper.

6. Refold the paper from top to bottom.

7. Using scissors, cut a slit along the center crease of the sheet from the folded edge to the creases made in step 4. Do not cut the entire sheet in half.

8. Fold the sheet of paper in half from left to right. While holding the bottom and top edges of the paper, push the bottom and top edges together so that the center collapses at the center slit. Fold the four flaps to form a four-page book.

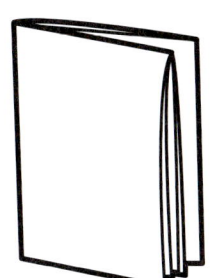

Layered Book

1. Lay one sheet of paper on top of another sheet. Slide the top sheet up so that 2 cm of the bottom sheet is showing.

2. Hold the two sheets together, fold down the top of the two sheets so that you see four 2 cm tabs along the bottom.

3. Using a stapler, staple the top of the FoldNote.

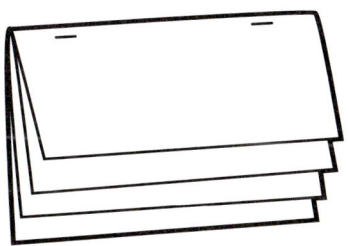

Key-Term Fold

1. Fold a sheet of lined notebook paper in half from left to right.

2. Using scissors, cut along every third line from the right edge of the paper to the center fold to make tabs.

Four-Corner Fold

1. Fold a sheet of paper in half from left to right. Then, unfold the paper.

2. Fold each side of the paper to the crease in the center of the paper.

3. Fold the paper in half from the top to the bottom. Then, unfold the paper.

4. Using scissors, cut the top flap creases made in step 3 to form four flaps.

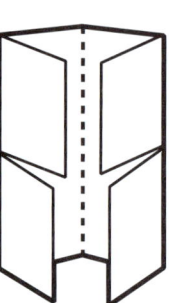

Three-Panel Flip Chart

1. Fold a piece of paper in half from the top to the bottom.

2. Fold the paper in thirds from side to side. Then, unfold the paper so that you can see the three sections.

3. From the top of the paper, cut along each of the vertical fold lines to the fold in the middle of the paper. You will now have three flaps.

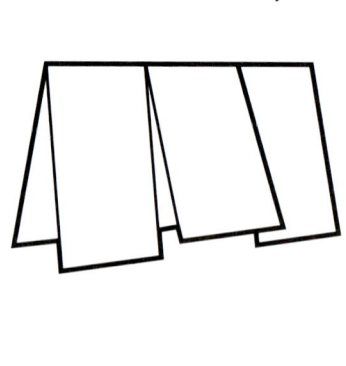

Table Fold

1. Fold a piece of paper in half from the top to the bottom. Then, fold the paper in half again.

2. Fold the paper in thirds from side to side.

3. Unfold the paper completely. Carefully trace the fold lines by using a pen or pencil.

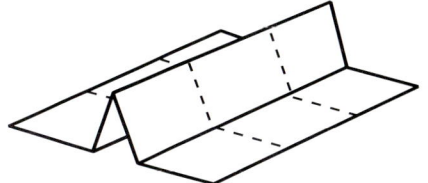

Two-Panel Flip Chart

1. Fold a piece of paper in half from the top to the bottom.

2. Fold the paper in half from side to side. Then, unfold the paper so that you can see the two sections.

3. From the top of the paper, cut along the vertical fold line to the fold in the middle of the paper. You will now have two flaps.

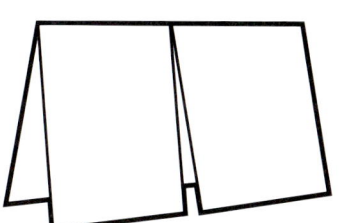

Tri-Fold

1. Fold a piece a paper in thirds from the top to the bottom.

2. Unfold the paper so that you can see the three sections. Then, turn the paper sideways so that the three sections form vertical columns.

3. Trace the fold lines by using a pen or pencil. Label the columns "Know," "Want," and "Learn."

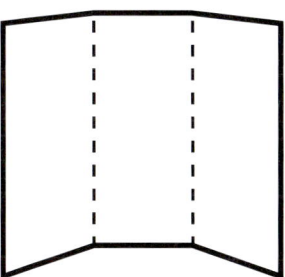

Appendix

Graphic Organizer Instructions

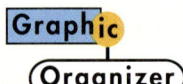 Have you ever wished that you could "draw out" the many concepts you learn in your science class? Sometimes, being able to *see* how concepts are related really helps you remember what you've learned. Graphic Organizers do just that! They give you a way to draw or map out concepts.

All you need to make a Graphic Organizer is a piece of paper and a pencil. Below you will find instructions for four different Graphic Organizers designed to help you organize the concepts you'll learn in this book.

Spider Map

1. Draw a diagram like the one shown. In the circle, write the main topic.

2. From the circle, draw legs to represent different categories of the main topic. You can have as many categories as you want.

3. From the category legs, draw horizontal lines. As you read the chapter, write details about each category on the horizontal lines.

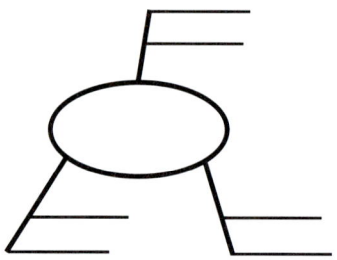

Comparison Table

1. Draw a chart like the one shown. Your chart can have as many columns and rows as you want.

2. In the top row, write the topics that you want to compare.

3. In the left column, write characteristics of the topics that you want to compare. As you read the chapter, fill in the characteristics for each topic in the appropriate boxes.

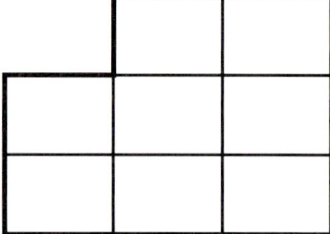

Chain-of-Events-Chart

1. Draw a box. In the box, write the first step of a process or the first event of a timeline.

2. Under the box, draw another box, and use an arrow to connect the two boxes. In the second box, write the next step of the process or the next event in the timeline.

3. Continue adding boxes until the process or timeline is finished.

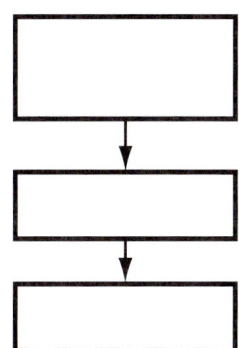

Concept Map

1. Draw a circle in the center of a piece of paper. Write the main idea of the chapter in the center of the circle.

2. From the circle, draw other circles. In those circles, write characteristics of the main idea. Draw arrows from the center circle to the circles that contain the characteristics.

3. From each circle that contains a characteristic, draw other circles. In those circles, write specific details about the characteristic. Draw arrows from each circle that contains a characteristic to the circles that contain specific details. You may draw as many circles as you want.

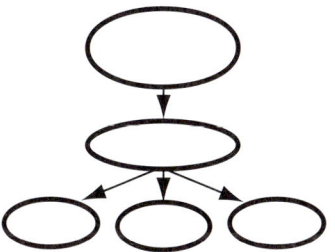

SI Measurement

The International System of Units, or SI, is the standard system of measurement used by many scientists. Using the same standards of measurement makes it easier for scientists to communicate with one another.

SI works by combining prefixes and base units. Each base unit can be used with different prefixes to define smaller and larger quantities. The table below lists common SI prefixes.

SI Prefixes			
Prefix	**Symbol**	**Factor**	**Example**
kilo-	k	1,000	kilogram, 1 kg = 1,000 g
hecto-	h	100	hectoliter, 1 hL = 100 L
deka-	da	10	dekameter, 1 dam = 10 m
		1	meter, liter, gram
deci-	d	0.1	decigram, 1 dg = 0.1 g
centi-	c	0.01	centimeter, 1 cm = 0.01 m
milli-	m	0.001	milliliter, 1 mL = 0.001 L
micro-	μ	0.000 001	micrometer, 1 μm = 0.000 001 m

SI Conversion Table		
SI units	**From SI to English**	**From English to SI**
Length		
kilometer (km) = 1,000 m	1 km = 0.621 mi	1 mi = 1.609 km
meter (m) = 100 cm	1 m = 3.281 ft	1 ft = 0.305 m
centimeter (cm) = 0.01 m	1 cm = 0.394 in.	1 in. = 2.540 cm
millimeter (mm) = 0.001 m	1 mm = 0.039 in.	
micrometer (μm) = 0.000 001 m		
nanometer (nm) = 0.000 000 001 m		
Area		
square kilometer (km^2) = 100 hectares	1 km^2 = 0.386 mi^2	1 mi^2 = 2.590 km^2
hectare (ha) = 10,000 m^2	1 ha = 2.471 acres	1 acre = 0.405 ha
square meter (m^2) = 10,000 cm^2	1 m^2 = 10.764 ft^2	1 ft^2 = 0.093 m^2
square centimeter (cm^2) = 100 mm^2	1 cm^2 = 0.155 in.^{2}	1 in.^{2} = 6.452 cm^2
Volume		
liter (L) = 1,000 mL = 1 dm^3	1 L = 1.057 fl qt	1 fl qt = 0.946 L
milliliter (mL) = 0.001 L = 1 cm^3	1 mL = 0.034 fl oz	1 fl oz = 29.574 mL
microliter (μL) = 0.000 001 L		
Mass		
kilogram (kg) = 1,000 g	1 kg = 2.205 lb	1 lb = 0.454 kg
gram (g) = 1,000 mg	1 g = 0.035 oz	1 oz = 28.350 g
milligram (mg) = 0.001 g		
microgram (μg) = 0.000 001 g		

Temperature Scales

Temperature can be expressed by using three different scales: Fahrenheit, Celsius, and Kelvin. The SI unit for temperature is the kelvin (K).

Although 0 K is much colder than 0°C, a change of 1 K is equal to a change of 1°C.

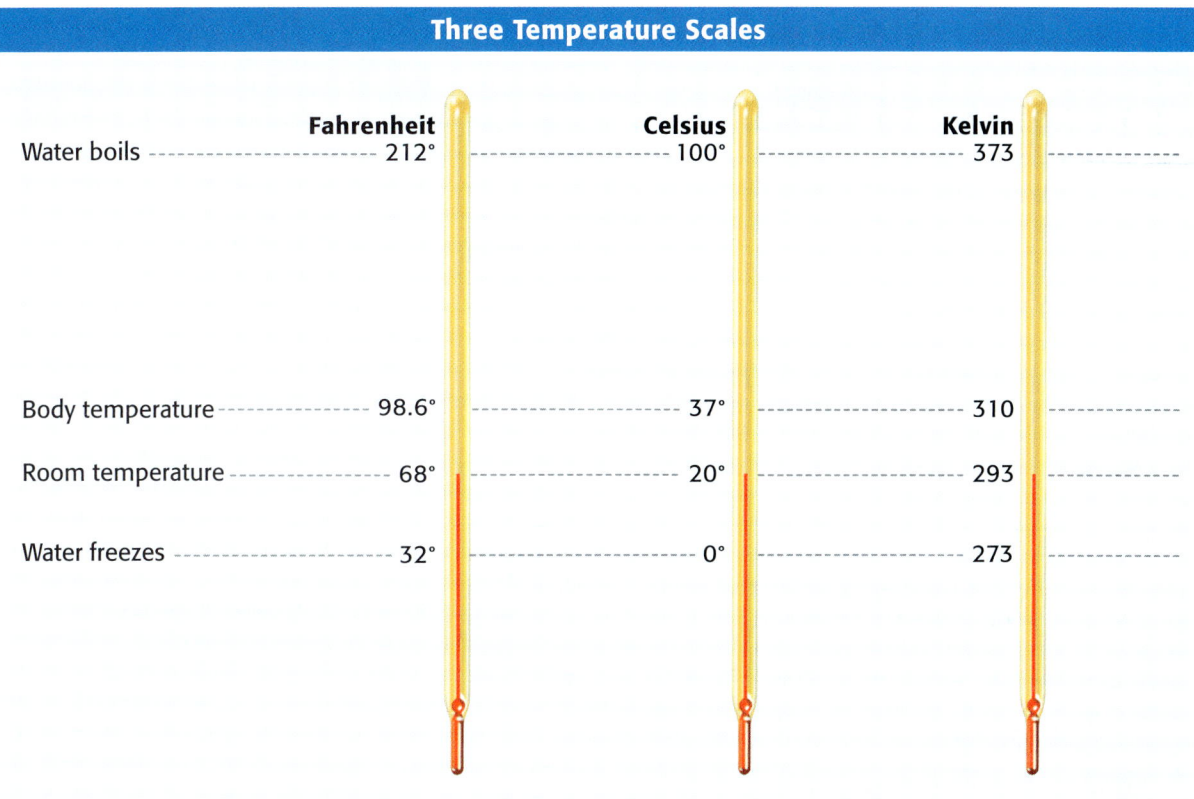

Three Temperature Scales

	Fahrenheit	Celsius	Kelvin
Water boils	212°	100°	373
Body temperature	98.6°	37°	310
Room temperature	68°	20°	293
Water freezes	32°	0°	273

Temperature Conversions Table		
To convert	**Use this equation:**	**Example**
Celsius to Fahrenheit °C → °F	$°F = \left(\dfrac{9}{5} \times °C\right) + 32$	Convert 45°C to °F. $°F = \left(\dfrac{9}{5} \times 45°C\right) + 32 = 113°F$
Fahrenheit to Celsius °F → °C	$°C = \dfrac{5}{9} \times (°F - 32)$	Convert 68°F to °C. $°C = \dfrac{5}{9} \times (68°F - 32) = 20°C$
Celsius to Kelvin °C → K	$K = °C + 273$	Convert 45°C to K. $K = 45°C + 273 = 318 \text{ K}$
Kelvin to Celsius K → °C	$°C = K - 273$	Convert 32 K to °C. $°C = 32K - 273 = -241°C$

Measuring Skills

Using a Graduated Cylinder

When using a graduated cylinder to measure volume, keep the following procedures in mind:

1 Place the cylinder on a flat, level surface before measuring liquid.

2 Move your head so that your eye is level with the surface of the liquid.

3 Read the mark closest to the liquid level. On glass graduated cylinders, read the mark closest to the center of the curve in the liquid's surface.

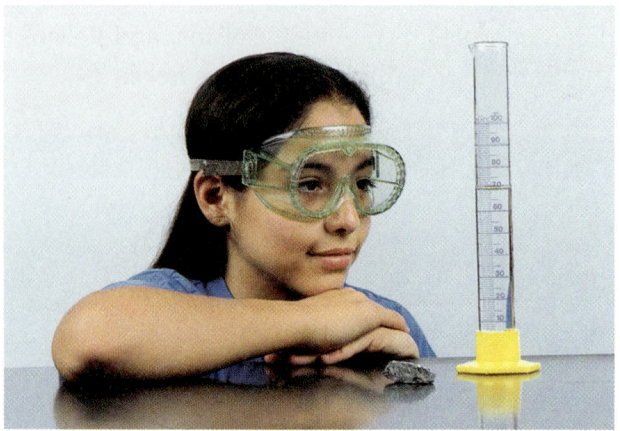

Using a Meterstick or Metric Ruler

When using a meterstick or metric ruler to measure length, keep the following procedures in mind:

1 Place the ruler firmly against the object that you are measuring.

2 Align one edge of the object exactly with the 0 end of the ruler.

3 Look at the other edge of the object to see which of the marks on the ruler is closest to that edge. (Note: Each small slash between the centimeters represents a millimeter, which is one-tenth of a centimeter.)

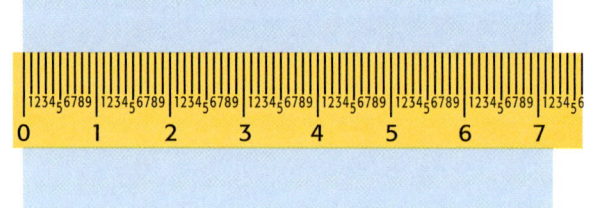

Using a Triple-Beam Balance

When using a triple-beam balance to measure mass, keep the following procedures in mind:

1 Make sure the balance is on a level surface.

2 Place all of the countermasses at 0. Adjust the balancing knob until the pointer rests at 0.

3 Place the object you wish to measure on the pan. **Caution:** Do not place hot objects or chemicals directly on the balance pan.

4 Move the largest countermass along the beam to the right until it is at the last notch that does not tip the balance. Follow the same procedure with the next-largest countermass. Then, move the smallest countermass until the pointer rests at 0.

5 Add the readings from the three beams together to determine the mass of the object.

6 When determining the mass of crystals or powders, first find the mass of a piece of filter paper. Then, add the crystals or powder to the paper, and remeasure. The actual mass of the crystals or powder is the total mass minus the mass of the paper. When finding the mass of liquids, first find the mass of the empty container. Then, find the combined mass of the liquid and container. The mass of the liquid is the total mass minus the mass of the container.

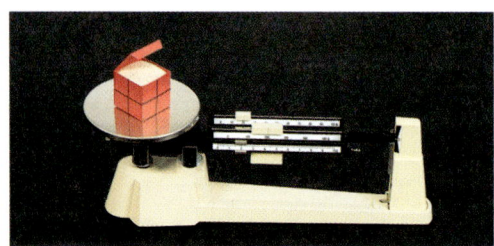

Scientific Methods

The ways in which scientists answer questions and solve problems are called **scientific methods.** The same steps are often used by scientists as they look for answers. However, there is more than one way to use these steps. Scientists may use all of the steps or just some of the steps during an investigation. They may even repeat some of the steps. The goal of using scientific methods is to come up with reliable answers and solutions.

Six Steps of Scientific Methods

1 Ask a Question

Good questions come from careful **observations.** You make observations by using your senses to gather information. Sometimes, you may use instruments, such as microscopes and telescopes, to extend the range of your senses. As you observe the natural world, you will discover that you have many more questions than answers. These questions drive investigations.

Questions beginning with *what, why, how,* and *when* are important in focusing an investigation. Here is an example of a question that could lead to an investigation.

Question: How does acid rain affect plant growth?

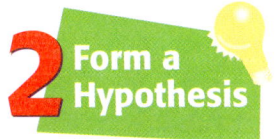
2 Form a Hypothesis

After you ask a question, you need to form a **hypothesis.** A hypothesis is a clear statement of what you expect the answer to your question to be. Your hypothesis will represent your best "educated guess" based on what you have observed and what you already know. A good hypothesis is testable. Otherwise, the investigation can go no further. Here is a hypothesis based on the question, "How does acid rain affect plant growth?"

Hypothesis: Acid rain slows plant growth.

The hypothesis can lead to predictions. A prediction is what you think the outcome of your experiment or data collection will be. Predictions are usually stated in an if-then format. Here is a sample prediction for the hypothesis that acid rain slows plant growth.

Prediction: If a plant is watered with only acid rain (which has a pH of 4), then the plant will grow at half its normal rate.

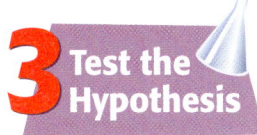
3 Test the Hypothesis

After you have formed a hypothesis and made a prediction, your hypothesis should be tested. One way to test a hypothesis is with a controlled experiment. A **controlled experiment** tests only one factor at a time. In an experiment to test the effect of acid rain on plant growth, the **control group** would be watered with normal rain water. The **experimental group** would be watered with acid rain. All of the plants should receive the same amount of sunlight and water each day. The air temperature should be the same for all groups. However, the acidity of the water will be a variable. In fact, any factor that is different from one group to another is a **variable.** If your hypothesis is correct, then the acidity of the water and plant growth are *dependant variables.* The amount a plant grows is dependent on the acidity of the water. However, the amount of water each plant receives and the amount of sunlight each plant receives are *independent variables.* Either of these factors could change without affecting the other factor.

Sometimes, the nature of an investigation makes a controlled experiment impossible. For example, the Earth's core is surrounded by thousands of meters of rock. Under such circumstances, a hypothesis may be tested by making detailed observations.

4 Analyze the Results

After you have completed your experiments, made your observations, and collected your data, you must analyze all the information you have gathered. Tables and graphs are often used in this step to organize the data.

5 Draw Conclusions

After analyzing your data, you can determine if your results support your hypothesis. If your hypothesis is supported, you (or others) might want to repeat the observations or experiments to verify your results. If your hypothesis is not supported by the data, you may have to check your procedure for errors. You may even have to reject your hypothesis and make a new one. If you cannot draw a conclusion from your results, you may have to try the investigation again or carry out further observations or experiments.

6 Communicate Results

After any scientific investigation, you should report your results. By preparing a written or oral report, you let others know what you have learned. They may repeat your investigation to see if they get the same results. Your report may even lead to another question and then to another investigation.

Scientific Methods in Action

Scientific methods contain loops in which several steps may be repeated over and over again. In some cases, certain steps are unnecessary. Thus, there is not a "straight line" of steps. For example, sometimes scientists find that testing one hypothesis raises new questions and new hypotheses to be tested. And sometimes, testing the hypothesis leads directly to a conclusion. Furthermore, the steps in scientific methods are not always used in the same order. Follow the steps in the diagram, and see how many different directions scientific methods can take you.

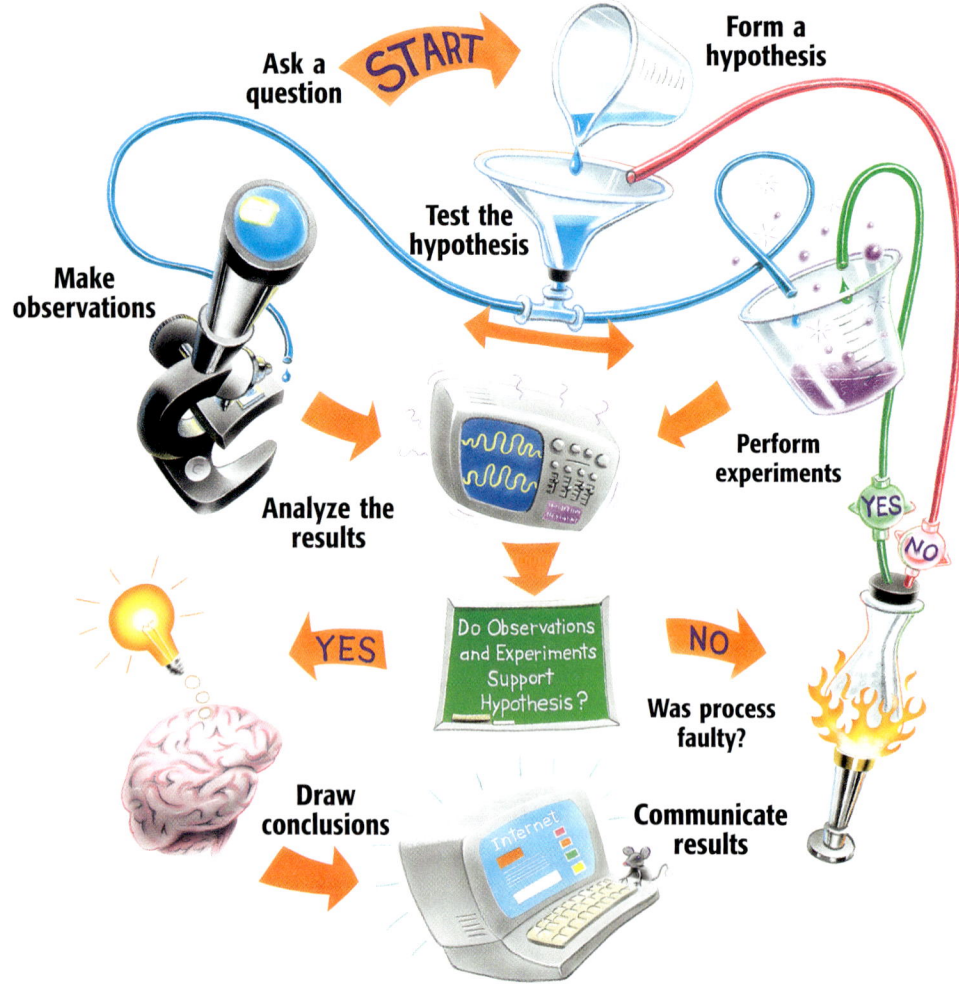

Making Charts and Graphs

Pie Charts

A pie chart shows how each group of data relates to all of the data. Each part of the circle forming the chart represents a category of the data. The entire circle represents all of the data. For example, a biologist studying a hardwood forest in Wisconsin found that there were five different types of trees. The data table at right summarizes the biologist's findings.

Wisconsin Hardwood Trees	
Type of tree	**Number found**
Oak	600
Maple	750
Beech	300
Birch	1,200
Hickory	150
Total	3,000

How to Make a Pie Chart

1 To make a pie chart of these data, first find the percentage of each type of tree. Divide the number of trees of each type by the total number of trees, and multiply by 100.

$$\frac{600 \text{ oak}}{3,000 \text{ trees}} \times 100 = 20\%$$

$$\frac{750 \text{ maple}}{3,000 \text{ trees}} \times 100 = 25\%$$

$$\frac{300 \text{ beech}}{3,000 \text{ trees}} \times 100 = 10\%$$

$$\frac{1,200 \text{ birch}}{3,000 \text{ trees}} \times 100 = 40\%$$

$$\frac{150 \text{ hickory}}{3,000 \text{ trees}} \times 100 = 5\%$$

2 Now, determine the size of the wedges that make up the pie chart. Multiply each percentage by 360°. Remember that a circle contains 360°.

$20\% \times 360° = 72°$ $25\% \times 360° = 90°$

$10\% \times 360° = 36°$ $40\% \times 360° = 144°$

$5\% \times 360° = 18°$

3 Check that the sum of the percentages is 100 and the sum of the degrees is 360.

$20\% + 25\% + 10\% + 40\% + 5\% = 100\%$

$72° + 90° + 36° + 144° + 18° = 360°$

4 Use a compass to draw a circle and mark the center of the circle.

5 Then, use a protractor to draw angles of 72°, 90°, 36°, 144°, and 18° in the circle.

6 Finally, label each part of the chart, and choose an appropriate title.

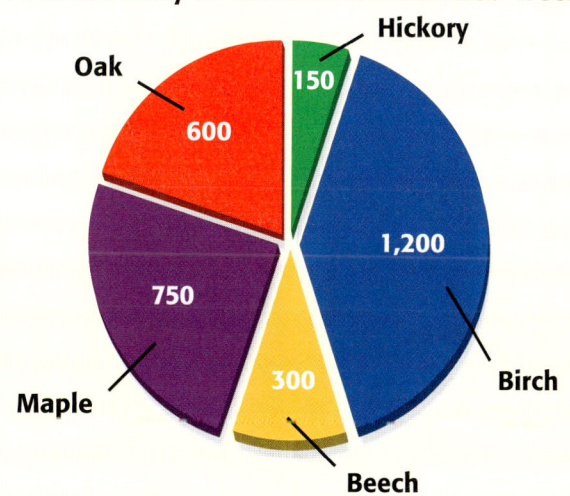

A Community of Wisconsin Hardwood Trees

Line Graphs

Line graphs are most often used to demonstrate continuous change. For example, Mr. Smith's students analyzed the population records for their hometown, Appleton, between 1900 and 2000. Examine the data at right.

Because the year and the population change, they are the *variables*. The population is determined by, or dependent on, the year. Therefore, the population is called the **dependent variable,** and the year is called the **independent variable.** Each set of data is called a **data pair.** To prepare a line graph, you must first organize data pairs into a table like the one at right.

Population of Appleton, 1900–2000	
Year	Population
1900	1,800
1920	2,500
1940	3,200
1960	3,900
1980	4,600
2000	5,300

How to Make a Line Graph

1. Place the independent variable along the horizontal (*x*) axis. Place the dependent variable along the vertical (*y*) axis.

2. Label the *x*-axis "Year" and the *y*-axis "Population." Look at your largest and smallest values for the population. For the *y*-axis, determine a scale that will provide enough space to show these values. You must use the same scale for the entire length of the axis. Next, find an appropriate scale for the *x*-axis.

3. Choose reasonable starting points for each axis.

4. Plot the data pairs as accurately as possible.

5. Choose a title that accurately represents the data.

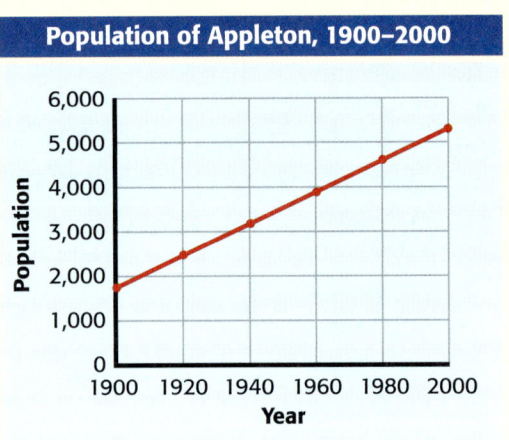

Population of Appleton, 1900–2000

How to Determine Slope

Slope is the ratio of the change in the *y*-value to the change in the *x*-value, or "rise over run."

1. Choose two points on the line graph. For example, the population of Appleton in 2000 was 5,300 people. Therefore, you can define point *a* as (2000, 5,300). In 1900, the population was 1,800 people. You can define point *b* as (1900, 1,800).

2. Find the change in the *y*-value. (*y* at point *a*) − (*y* at point *b*) = 5,300 people − 1,800 people = 3,500 people

3. Find the change in the *x*-value. (*x* at point *a*) − (*x* at point *b*) = 2000 − 1900 = 100 years

4. Calculate the slope of the graph by dividing the change in *y* by the change in *x*.

$$slope = \frac{change\ in\ y}{change\ in\ x}$$

$$slope = \frac{3{,}500\ people}{100\ years}$$

$$slope = 35\ people\ per\ year$$

In this example, the population in Appleton increased by a fixed amount each year. The graph of these data is a straight line. Therefore, the relationship is **linear.** When the graph of a set of data is not a straight line, the relationship is **nonlinear.**

Using Algebra to Determine Slope

The equation in step 4 may also be arranged to be

$$y = kx$$

where y represents the change in the y-value, k represents the slope, and x represents the change in the x-value.

$$slope = \frac{change\ in\ y}{change\ in\ x}$$

$$k = \frac{y}{x}$$

$$k \times x = \frac{y \times x}{x}$$

$$kx = y$$

Bar Graphs

Bar graphs are used to demonstrate change that is not continuous. These graphs can be used to indicate trends when the data cover a long period of time. A meteorologist gathered the precipitation data shown here for Hartford, Connecticut, for April 1–15, 1996, and used a bar graph to represent the data.

Precipitation in Hartford, Connecticut April 1–15, 1996			
Date	Precipitation (cm)	Date	Precipitation (cm)
April 1	0.5	April 9	0.25
April 2	1.25	April 10	0.0
April 3	0.0	April 11	1.0
April 4	0.0	April 12	0.0
April 5	0.0	April 13	0.25
April 6	0.0	April 14	0.0
April 7	0.0	April 15	6.50
April 8	1.75		

How to Make a Bar Graph

1 Use an appropriate scale and a reasonable starting point for each axis.

2 Label the axes, and plot the data.

3 Choose a title that accurately represents the data.

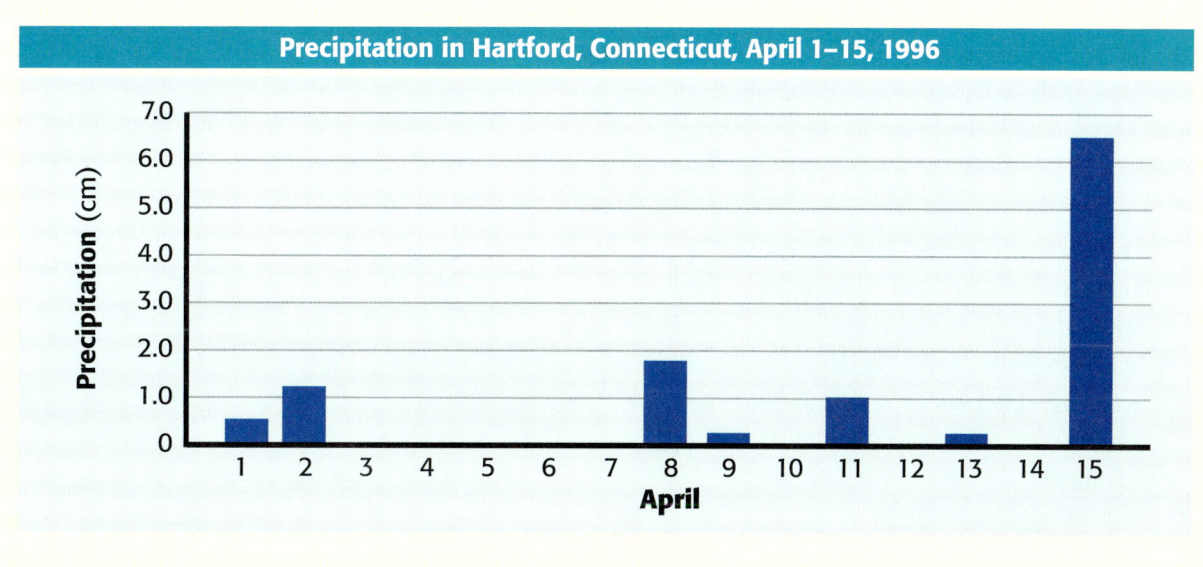

Math Refresher

Science requires an understanding of many math concepts. The following pages will help you review some important math skills.

Averages

An **average,** or **mean,** simplifies a set of numbers into a single number that *approximates* the value of the set.

> **Example:** Find the average of the following set of numbers: 5, 4, 7, and 8.

Step 1: Find the sum.

$$5 + 4 + 7 + 8 = 24$$

Step 2: Divide the sum by the number of numbers in your set. Because there are four numbers in this example, divide the sum by 4.

$$\frac{24}{4} = 6$$

The average, or mean, is **6.**

Ratios

A **ratio** is a comparison between numbers, and it is usually written as a fraction.

> **Example:** Find the ratio of thermometers to students if you have 36 thermometers and 48 students in your class.

Step 1: Make the ratio.

$$\frac{36 \text{ thermometers}}{48 \text{ students}}$$

Step 2: Reduce the fraction to its simplest form.

$$\frac{36}{48} = \frac{36 \div 12}{48 \div 12} = \frac{3}{4}$$

The ratio of thermometers to students is **3 to 4,** or $\frac{3}{4}$. The ratio may also be written in the form 3:4.

Proportions

A **proportion** is an equation that states that two ratios are equal.

$$\frac{3}{1} = \frac{12}{4}$$

To solve a proportion, first multiply across the equal sign. This is called *cross-multiplication.* If you know three of the quantities in a proportion, you can use cross-multiplication to find the fourth.

> **Example:** Imagine that you are making a scale model of the solar system for your science project. The diameter of Jupiter is 11.2 times the diameter of the Earth. If you are using a plastic-foam ball that has a diameter of 2 cm to represent the Earth, what must the diameter of the ball representing Jupiter be?
>
> $$\frac{11.2}{1} = \frac{x}{2 \text{ cm}}$$

Step 1: Cross-multiply.

$$\frac{11.2}{1} \times \frac{x}{2}$$

$$11.2 \times 2 = x \times 1$$

Step 2: Multiply.

$$22.4 = x \times 1$$

Step 3: Isolate the variable by dividing both sides by 1.

$$x = \frac{22.4}{1}$$

$$x = 22.4 \text{ cm}$$

You will need to use a ball that has a diameter of **22.4** cm to represent Jupiter.

Percentages

A **percentage** is a ratio of a given number to 100.

Example: What is 85% of 40?

Step 1: Rewrite the percentage by moving the decimal point two places to the left.

0.85

Step 2: Multiply the decimal by the number that you are calculating the percentage of.

$0.85 \times 40 = 34$

85% of 40 is **34.**

Decimals

To **add** or **subtract decimals,** line up the digits vertically so that the decimal points line up. Then, add or subtract the columns from right to left. Carry or borrow numbers as necessary.

Example: Add the following numbers: 3.1415 and 2.96.

Step 1: Line up the digits vertically so that the decimal points line up.

$$\begin{array}{r} 3.1415 \\ + 2.96 \\ \hline \end{array}$$

Step 2: Add the columns from right to left, and carry when necessary.

$$\begin{array}{r} {}^{1}\ {}^{1} \\ 3.1415 \\ + 2.96 \\ \hline 6.1015 \end{array}$$

The sum is **6.1015.**

Fractions

Numbers tell you how many; **fractions** tell you *how much of a whole*.

Example: Your class has 24 plants. Your teacher instructs you to put 5 plants in a shady spot. What fraction of the plants in your class will you put in a shady spot?

Step 1: In the denominator, write the total number of parts in the whole.

$$\frac{?}{24}$$

Step 2: In the numerator, write the number of parts of the whole that are being considered.

$$\frac{5}{24}$$

So, $\frac{5}{24}$ of the plants will be in the shade.

Reducing Fractions

It is usually best to express a fraction in its simplest form. Expressing a fraction in its simplest form is called *reducing* a fraction.

Example: Reduce the fraction $\frac{30}{45}$ to its simplest form.

Step 1: Find the largest whole number that will divide evenly into both the numerator and denominator. This number is called the *greatest common factor* (GCF).

Factors of the numerator 30:

1, 2, 3, 5, 6, 10, **15,** 30

Factors of the denominator 45:

1, 3, 5, 9, **15,** 45

Step 2: Divide both the numerator and the denominator by the GCF, which in this case is 15.

$$\frac{30}{45} = \frac{30 \div 15}{45 \div 15} = \frac{2}{3}$$

Thus, $\frac{30}{45}$ reduced to its simplest form is $\frac{2}{3}$.

Adding and Subtracting Fractions

To **add** or **subtract fractions** that have the **same denominator,** simply add or subtract the numerators.

Examples:

$$\frac{3}{5} + \frac{1}{5} = ? \text{ and } \frac{3}{4} - \frac{1}{4} = ?$$

Step 1: Add or subtract the numerators.

$$\frac{3}{5} + \frac{1}{5} = \frac{4}{\_} \text{ and } \frac{3}{4} - \frac{1}{4} = \frac{2}{\_}$$

Step 2: Write the sum or difference over the denominator.

$$\frac{3}{5} + \frac{1}{5} = \frac{4}{5} \text{ and } \frac{3}{4} - \frac{1}{4} = \frac{2}{4}$$

Step 3: If necessary, reduce the fraction to its simplest form.

$\frac{4}{5}$ cannot be reduced, and $\frac{2}{4} = \frac{1}{2}$.

To **add** or **subtract fractions** that have **different denominators,** first find the least common denominator (LCD).

Examples:

$$\frac{1}{2} + \frac{1}{6} = ? \text{ and } \frac{3}{4} - \frac{2}{3} = ?$$

Step 1: Write the equivalent fractions that have a common denominator.

$$\frac{3}{6} + \frac{1}{6} = ? \text{ and } \frac{9}{12} - \frac{8}{12} = ?$$

Step 2: Add or subtract the fractions.

$$\frac{3}{6} + \frac{1}{6} = \frac{4}{6} \text{ and } \frac{9}{12} - \frac{8}{12} = \frac{1}{12}$$

Step 3: If necessary, reduce the fraction to its simplest form.

The fraction $\frac{4}{6} = \frac{2}{3}$, and $\frac{1}{12}$ cannot be reduced.

Multiplying Fractions

To **multiply fractions,** multiply the numerators and the denominators together, and then reduce the fraction to its simplest form.

Example:

$$\frac{5}{9} \times \frac{7}{10} = ?$$

Step 1: Multiply the numerators and denominators.

$$\frac{5}{9} \times \frac{7}{10} = \frac{5 \times 7}{9 \times 10} = \frac{35}{90}$$

Step 2: Reduce the fraction.

$$\frac{35}{90} = \frac{35 \div 5}{90 \div 5} = \frac{7}{18}$$

Dividing Fractions

To **divide fractions,** first rewrite the divisor (the number you divide by) upside down. This number is called the *reciprocal* of the divisor. Then multiply and reduce if necessary.

Example:

$$\frac{5}{8} \div \frac{3}{2} = ?$$

Step 1: Rewrite the divisor as its reciprocal.

$$\frac{3}{2} \rightarrow \frac{2}{3}$$

Step 2: Multiply the fractions.

$$\frac{5}{8} \times \frac{2}{3} = \frac{5 \times 2}{8 \times 3} = \frac{10}{24}$$

Step 3: Reduce the fraction.

$$\frac{10}{24} = \frac{10 \div 2}{24 \div 2} = \frac{5}{12}$$

Scientific Notation

Scientific notation is a short way of representing very large and very small numbers without writing all of the place-holding zeros.

Example: Write 653,000,000 in scientific notation.

Step 1: Write the number without the place-holding zeros.

653

Step 2: Place the decimal point after the first digit.

6.53

Step 3: Find the exponent by counting the number of places that you moved the decimal point.

6.53000000

The decimal point was moved eight places to the left. Therefore, the exponent of 10 is positive 8. If you had moved the decimal point to the right, the exponent would be negative.

Step 4: Write the number in scientific notation.

$$\textbf{6.53} \times \textbf{10}^8$$

Area

Area is the number of square units needed to cover the surface of an object.

Formulas:

$area\ of\ a\ square = side \times side$
$area\ of\ a\ rectangle = length \times width$
$area\ of\ a\ triangle = \frac{1}{2} \times base \times height$

Examples: Find the areas.

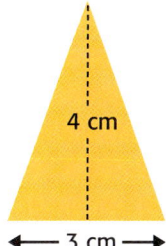

Triangle

$area = \frac{1}{2} \times base \times height$

$area = \frac{1}{2} \times 3\ cm \times 4\ cm$

$area = \textbf{6 cm}^2$

Rectangle

$area = length \times width$
$area = 6\ cm \times 3\ cm$
$area = \textbf{18 cm}^2$

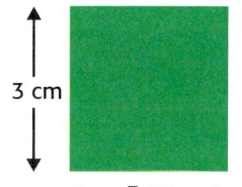

Square

$area = side \times side$
$area = 3\ cm \times 3\ cm$
$area = \textbf{9 cm}^2$

Volume

Volume is the amount of space that something occupies.

Formulas:

$volume\ of\ a\ cube =$
$side \times side \times side$

$volume\ of\ a\ prism =$
$area\ of\ base \times height$

Examples:

Find the volume of the solids.

Cube

$volume = side \times side \times side$
$volume = 4\ cm \times 4\ cm \times 4\ cm$
$volume = \textbf{64 cm}^3$

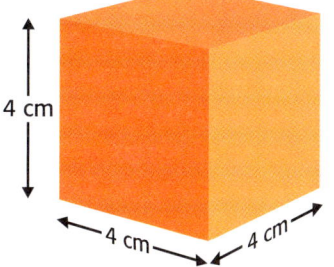

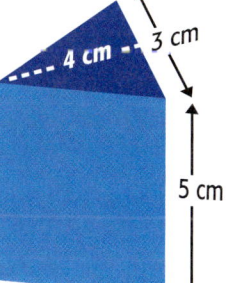

Prism

$volume = area\ of\ base \times height$
$volume = (area\ of\ triangle) \times height$
$volume = (\frac{1}{2} \times 3\ cm \times 4\ cm) \times 5\ cm$
$volume = 6\ cm^2 \times 5\ cm$
$volume = \textbf{30 cm}^3$

Periodic Table of the Elements

Each square on the table includes an element's name, chemical symbol, atomic number, and atomic mass.

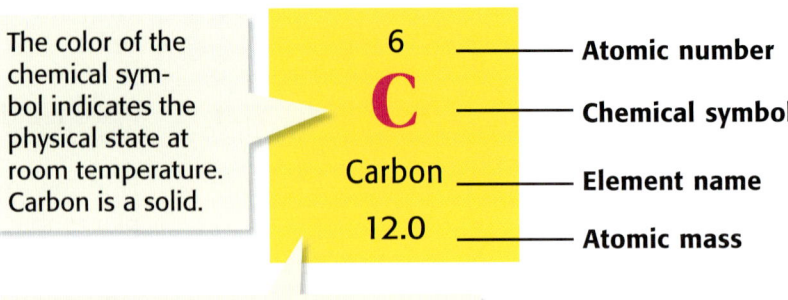

The color of the chemical symbol indicates the physical state at room temperature. Carbon is a solid.

6
C
Carbon
12.0

Atomic number — 6
Chemical symbol — C
Element name — Carbon
Atomic mass — 12.0

The background color indicates the type of element. Carbon is a nonmetal.

Period 1

1
H
Hydrogen
1.0

Background

Metals	(light blue)
Metalloids	(light green)
Nonmetals	(yellow)

Chemical symbol

Solid	(red)
Liquid	(blue)
Gas	(green)

	Group 1	Group 2	Group 3	Group 4	Group 5	Group 6	Group 7	Group 8	Group 9
Period 2	3 **Li** Lithium 6.9	4 **Be** Beryllium 9.0							
Period 3	11 **Na** Sodium 23.0	12 **Mg** Magnesium 24.3							
Period 4	19 **K** Potassium 39.1	20 **Ca** Calcium 40.1	21 **Sc** Scandium 45.0	22 **Ti** Titanium 47.9	23 **V** Vanadium 50.9	24 **Cr** Chromium 52.0	25 **Mn** Manganese 54.9	26 **Fe** Iron 55.8	27 **Co** Cobalt 58.9
Period 5	37 **Rb** Rubidium 85.5	38 **Sr** Strontium 87.6	39 **Y** Yttrium 88.9	40 **Zr** Zirconium 91.2	41 **Nb** Niobium 92.9	42 **Mo** Molybdenum 95.9	43 **Tc** Technetium (98)	44 **Ru** Ruthenium 101.1	45 **Rh** Rhodium 102.9
Period 6	55 **Cs** Cesium 132.9	56 **Ba** Barium 137.3	57 **La** Lanthanum 138.9	72 **Hf** Hafnium 178.5	73 **Ta** Tantalum 180.9	74 **W** Tungsten 183.8	75 **Re** Rhenium 186.2	76 **Os** Osmium 190.2	77 **Ir** Iridium 192.2
Period 7	87 **Fr** Francium (223)	88 **Ra** Radium (226)	89 **Ac** Actinium (227)	104 **Rf** Rutherfordium (261)	105 **Db** Dubnium (262)	106 **Sg** Seaborgium (263)	107 **Bh** Bohrium (264)	108 **Hs** Hassium (265)[†]	109 **Mt** Meitnerium (268)[†]

† Estimated from currently available IUPAC data.

A row of elements is called a *period*.

A column of elements is called a *group* or *family*.

Values in parentheses are of the most stable isotope of the element.

These elements are placed below the table to allow the table to be narrower.

Lanthanides	58 **Ce** Cerium 140.1	59 **Pr** Praseodymium 140.9	60 **Nd** Neodymium 144.2	61 **Pm** Promethium (145)	62 **Sm** Samarium 150.4
Actinides	90 **Th** Thorium 232.0	91 **Pa** Protactinium 231.0	92 **U** Uranium 238.0	93 **Np** Neptunium (237)	94 **Pu** Plutonium (244)

Appendix

Topic: **Periodic Table**
Go To: **go.hrw.com**
Keyword: **HN0 PERIODIC**
Visit the HRW Web site for
updates on the periodic table.

This zigzag line reminds you where the metals, nonmetals, and metalloids are.

Group 18

			Group 13	Group 14	Group 15	Group 16	Group 17	2 **He** Helium 4.0
			5 **B** Boron 10.8	6 **C** Carbon 12.0	7 **N** Nitrogen 14.0	8 **O** Oxygen 16.0	9 **F** Fluorine 19.0	10 **Ne** Neon 20.2
Group 10	Group 11	Group 12	13 **Al** Aluminum 27.0	14 **Si** Silicon 28.1	15 **P** Phosphorus 31.0	16 **S** Sulfur 32.1	17 **Cl** Chlorine 35.5	18 **Ar** Argon 39.9
28 **Ni** Nickel 58.7	29 **Cu** Copper 63.5	30 **Zn** Zinc 65.4	31 **Ga** Gallium 69.7	32 **Ge** Germanium 72.6	33 **As** Arsenic 74.9	34 **Se** Selenium 79.0	35 **Br** Bromine 79.9	36 **Kr** Krypton 83.8
46 **Pd** Palladium 106.4	47 **Ag** Silver 107.9	48 **Cd** Cadmium 112.4	49 **In** Indium 114.8	50 **Sn** Tin 118.7	51 **Sb** Antimony 121.8	52 **Te** Tellurium 127.6	53 **I** Iodine 126.9	54 **Xe** Xenon 131.3
78 **Pt** Platinum 195.1	79 **Au** Gold 197.0	80 **Hg** Mercury 200.6	81 **Tl** Thallium 204.4	82 **Pb** Lead 207.2	83 **Bi** Bismuth 209.0	84 **Po** Polonium (209)	85 **At** Astatine (210)	86 **Rn** Radon (222)
110 **Ds** Darmstadtium (269)†	111 **Uuu** Unununium (272)†	112 **Uub** Ununbium (277)†		114 **Uuq** Ununquadium (285)†				

The names and three-letter symbols of elements are temporary. They are based on the atomic numbers of the elements. Official names and symbols will be approved by an international committee of scientists.

63 **Eu** Europium 152.0	64 **Gd** Gadolinium 157.2	65 **Tb** Terbium 158.9	66 **Dy** Dysprosium 162.5	67 **Ho** Holmium 164.9	68 **Er** Erbium 167.3	69 **Tm** Thulium 168.9	70 **Yb** Ytterbium 173.0	71 **Lu** Lutetium 175.0
95 **Am** Americium (243)	96 **Cm** Curium (247)	97 **Bk** Berkelium (247)	98 **Cf** Californium (251)	99 **Es** Einsteinium (252)	100 **Fm** Fermium (257)	101 **Md** Mendelevium (258)	102 **No** Nobelium (259)	103 **Lr** Lawrencium (262)

Appendix

Physical Science Refresher

Atoms and Elements

Every object in the universe is made up of particles of some kind of matter. **Matter** is anything that takes up space and has mass. All matter is made up of elements. An **element** is a substance that cannot be separated into simpler components by ordinary chemical means. This is because each element consists of only one kind of atom. An **atom** is the smallest unit of an element that has all of the properties of that element.

Atomic Structure

Atoms are made up of small particles called subatomic particles. The three major types of subatomic particles are **electrons, protons, and neutrons.** Electrons have a negative electric charge, protons have a positive charge, and neutrons have no electric charge. The protons and neutrons are packed close to one another to form the **nucleus.** The protons give the nucleus a positive charge. Electrons are most likely to be found in regions around the nucleus called **electron clouds.** The negatively charged electrons are attracted to the positively charged nucleus. An atom may have several energy levels in which electrons are located.

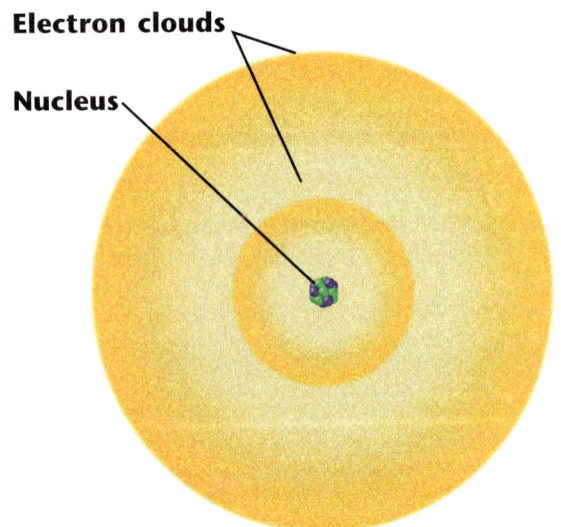

Electron clouds

Nucleus

Atomic Number

To help in the identification of elements, scientists have assigned an **atomic number** to each kind of atom. The atomic number is the number of protons in the atom. Atoms with the same number of protons are all the same kind of element. In an uncharged, or electrically neutral, atom there are an equal number of protons and electrons. Therefore, the atomic number equals the number of electrons in an uncharged atom. The number of neutrons, however, can vary for a given element. Atoms of the same element that have different numbers of neutrons are called **isotopes.**

Periodic Table of the Elements

In the periodic table, the elements are arranged from left to right in order of increasing atomic number. Each element in the table is in a separate box. An uncharged atom of each element has one more electron and one more proton than an uncharged atom of the element to its left. Each horizontal row of the table is called a **period.** Changes in chemical properties of elements across a period correspond to changes in the electron arrangements of their atoms. Each vertical column of the table, known as a **group,** lists elements with similar properties. The elements in a group have similar chemical properties because their atoms have the same number of electrons in their outer energy level. For example, the elements helium, neon, argon, krypton, xenon, and radon all have similar properties and are known as the noble gases.

Molecules and Compounds

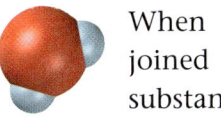 When two or more elements are joined chemically, the resulting substance is called a **compound.** A compound is a new substance with properties different from those of the elements that compose it. For example, water, H_2O, is a compound formed when hydrogen (H) and oxygen (O) combine. The smallest complete unit of a compound that has the properties of that compound is called a **molecule.** A chemical formula indicates the elements in a compound. It also indicates the relative number of atoms of each element present. The chemical formula for water is H_2O, which indicates that each water molecule consists of two atoms of hydrogen and one atom of oxygen. The subscript number after the symbol for an element indicates how many atoms of that element are in a single molecule of the compound.

Acids, Bases, and pH

An ion is an atom or group of atoms that has an electric charge because it has lost or gained one or more electrons. When an acid, such as hydrochloric acid, HCl, is mixed with water, it separates into ions. An **acid** is a compound that produces hydrogen ions, H+, in water. The hydrogen ions then combine with a water molecule to form a hydronium ion, H_3O^+. A **base,** on the other hand, is a substance that produces hydroxide ions, OH⁻, in water.

To determine whether a solution is acidic or basic, scientists use pH. The **pH** is a measure of the hydronium ion concentration in a solution. The pH scale ranges from 0 to 14. The middle point, pH = 7, is neutral, neither acidic nor basic. Acids have a pH less than 7; bases have a pH greater than 7. The lower the number is, the more acidic the solution. The higher the number is, the more basic the solution.

Chemical Equations

A chemical reaction occurs when a chemical change takes place. (In a chemical change, new substances with new properties are formed.) A chemical equation is a useful way of describing a chemical reaction by means of chemical formulas. The equation indicates what substances react and what the products are. For example, when carbon and oxygen combine, they can form carbon dioxide. The equation for the reaction is as follows: $C + O_2 \rightarrow CO_2$.

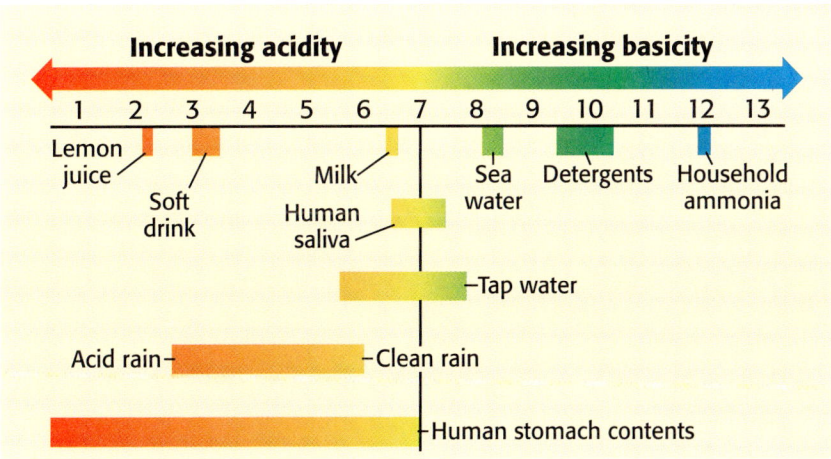

Physical Science Laws and Equations

Law of Conservation of Energy

The law of conservation of energy states that energy can be neither created nor destroyed.

The total amount of energy in a closed system is always the same. Energy can be changed from one form to another, but all of the different forms of energy in a system always add up to the same total amount of energy no matter how many energy conversions occur.

Law of Universal Gravitation

The law of universal gravitation states that all objects in the universe attract each other by a force called *gravity*. The size of the force depends on the masses of the objects and the distance between objects.

The first part of the law explains why a bowling ball is much harder to lift than a table-tennis ball. Because the bowling ball has a much larger mass than the table-tennis ball does, the amount of gravity between the Earth and the bowling ball is greater than the amount of gravity between the Earth and the table-tennis ball.

The second part of the law explains why a satellite can remain in orbit around the Earth. The satellite is carefully placed at a distance great enough to prevent the Earth's gravity from immediately pulling the satellite down but small enough to prevent the satellite from completely escaping the Earth's gravity and wandering off into space.

Newton's Laws of Motion

Newton's first law of motion states that an object at rest remains at rest and an object in motion remains in motion at constant speed and in a straight line unless acted on by an unbalanced force.

The first part of the law explains why a football will remain on a tee until it is kicked off or until a gust of wind blows it off.

The second part of the law explains why a bike rider will continue moving forward after the bike comes to an abrupt stop. Gravity and the friction of the sidewalk will eventually stop the rider.

Newton's second law of motion states that the acceleration of an object depends on the mass of the object and the amount of force applied.

The first part of the law explains why the acceleration of a 4 kg bowling ball will be greater than the acceleration of a 6 kg bowling ball if the same force is applied to both.

The second part of the law explains why the acceleration of a bowling ball will be larger if a larger force is applied to the bowling ball.

The relationship of acceleration (*a*) to mass (*m*) and force (*F*) can be expressed mathematically by the following equation:

$$acceleration = \frac{force}{mass}, \text{ or } a = \frac{F}{m}$$

This equation is often rearranged to the form

$$force = mass \times acceleration$$
$$\text{or}$$
$$F = m \times a$$

Newton's third law of motion states that whenever one object exerts a force on a second object, the second object exerts an equal and opposite force on the first.

This law explains that a runner is able to move forward because of the equal and opposite force that the ground exerts on the runner's foot after each step.

Useful Equations

Average speed

$$average\ speed = \frac{total\ distance}{total\ time}$$

Example: A bicycle messenger traveled a distance of 136 km in 8 h. What was the messenger's average speed?

$$\frac{136\ km}{8\ h} = 17\ km/h$$

The messenger's average speed was **17 km/h.**

Average acceleration

$$\frac{average}{acceleration} = \frac{final\ velocity - starting\ velocity}{time\ it\ takes\ to\ change\ velocity}$$

Example: Calculate the average acceleration of an Olympic 100 m dash sprinter who reaches a velocity of 20 m/s south at the finish line. The race was in a straight line and lasted 10 s.

$$\frac{20\ m/s - 0\ m/s}{10\ s} = 2\ m/s/s$$

The sprinter's average acceleration is **2 m/s/s south.**

Net force

Forces in the Same Direction

When forces are in the same direction, add the forces together to determine the net force.

Example: Calculate the net force on a stalled car that is being pushed by two people. One person is pushing with a force of 13 N northwest, and the other person is pushing with a force of 8 N in the same direction.

$$13\ N + 8\ N = 21\ N$$

The net force is **21 N northwest.**

Forces in Opposite Directions

When forces are in opposite directions, subtract the smaller force from the larger force to determine the net force. The net force will be in the direction of the larger force.

Net force (continued)

Example: Calculate the net force on a rope that is being pulled on each end. One person is pulling on one end of the rope with a force of 12 N south. Another person is pulling on the opposite end of the rope with a force of 7 N north.

$$12\ N - 7\ N = 5\ N$$

The net force is **5 N south.**

Density

$$density = \frac{mass}{volume}$$

Example: Calculate the density of a sponge that has a mass of 10 g and a volume of 40 cm^3.

$$\frac{10\ g}{40\ cm^3} = \frac{0.25g}{cm^3}$$

The density of the sponge is **0.25 g/cm^3.**

Pressure

Pressure is the force exerted over a given area. The SI unit for pressure is the pascal, whose symbol is Pa.

$$pressure = \frac{force}{area}$$

Example: Calculate the pressure of the air in a soccer ball if the air exerts a force of 10 N over an area of 0.5 m^2.

$$pressure = \frac{10\ N}{0.5\ m^2} = \frac{20\ N}{m^2} = 20\ Pa$$

The pressure of the air inside the soccer ball is **20 Pa.**

Concentration

$$concentration = \frac{mass\ of\ solute}{volume\ of\ solvent}$$

Example: Calculate the concentration of a solution in which 10 g of sugar is dissolved in 125 mL of water.

$$\frac{10\ g\ of\ sugar}{125\ mL\ of\ water} = \frac{0.08\ g}{mL}$$

The concentration of this solution is **0.08 g/mL.**

Properties of Common Minerals

Silicate Minerals

Mineral	Color	Luster	Streak	Hardness
Beryl	deep green, pink, white, bluish green, or yellow	vitreous	white	7.5–8
Chlorite	green	vitreous to pearly	pale green	2–2.5
Garnet	green, red, brown, black	vitreous	white	6.5–7.5
Hornblende	dark green, brown, or black	vitreous	none	5–6
Muscovite	colorless, silvery white, or brown	vitreous or pearly	white	2–2.5
Olivine	olive green, yellow	vitreous	white or none	6.5–7
Orthoclase	colorless, white, pink, or other colors	vitreous	white or none	6
Plagioclase	colorless, white, yellow, pink, green	vitreous	white	6
Quartz	colorless or white; any color when not pure	vitreous or waxy	white or none	7

Nonsilicate Minerals

Mineral	Color	Luster	Streak	Hardness
Native Elements				
Copper	copper-red	metallic	copper-red	2.5–3
Diamond	pale yellow or colorless	adamantine	none	10
Graphite	black to gray	submetallic	black	1–2
Carbonates				
Aragonite	colorless, white, or pale yellow	vitreous	white	3.5–4
Calcite	colorless or white to tan	vitreous	white	3
Halides				
Fluorite	light green, yellow, purple, bluish green, or other colors	vitreous	none	4
Halite	white	vitreous	white	2.0–2.5
Oxides				
Hematite	reddish brown to black	metallic to earthy	dark red to red-brown	5.6–6.5
Magnetite	iron-black	metallic	black	5.5–6.5
Sulfates				
Anhydrite	colorless, bluish, or violet	vitreous to pearly	white	3–3.5
Gypsum	white, pink, gray, or colorless	vitreous, pearly, or silky	white	2.0
Sulfides				
Galena	lead-gray	metallic	lead-gray to black	2.5–2.8
Pyrite	brassy yellow	metallic	greenish, brownish, or black	6–6.5

Density (g/cm^3)	Cleavage, Fracture, Special Properties	Common Uses
2.6–2.8	1 cleavage direction; irregular fracture; some varieties fluoresce in ultraviolet light	gemstones, ore of the metal beryllium
2.6–3.3	1 cleavage direction; irregular fracture	
4.2	no cleavage; conchoidal to splintery fracture	gemstones, abrasives
3.0–3.4	2 cleavage directions; hackly to splintery fracture	
2.7–3	1 cleavage direction; irregular fracture	electrical insulation, wallpaper, fireproofing material, lubricant
3.2–3.3	no cleavage; conchoidal fracture	gemstones, casting
2.6	2 cleavage directions; irregular fracture	porcelain
2.6–2.7	2 cleavage directions; irregular fracture	ceramics
2.6	no cleavage; conchoidal fracture	gemstones, concrete, glass, porcelain, sandpaper, lenses
8.9	no cleavage; hackly fracture	wiring, brass, bronze, coins
3.5	4 cleavage directions; irregular to conchoidal fracture	gemstones, drilling
2.3	1 cleavage direction; irregular fracture	pencils, paints, lubricants, batteries
2.95	2 cleavage directions; irregular fracture; reacts with hydrochloric acid	no important industrial uses
2.7	3 cleavage directions; irregular fracture; reacts with weak acid; double refraction	cements, soil conditioner, whitewash, construction materials
3.0–3.3	4 cleavage directions; irregular fracture; some varieties fluoresce	hydrofluoric acid, steel, glass, fiberglass, pottery, enamel
2.1–2.2	3 cleavage directions; splintery to conchoidal fracture; salty taste	tanning hides, salting icy roads, food preservation
5.2–5.3	no cleavage; splintery fracture; magnetic when heated	iron ore for steel, pigments
5.2	no cleavage; splintery fracture; magnetic	iron ore
3.0	3 cleavage directions; conchoidal to splintery fracture	soil conditioner, sulfuric acid
2.3	3 cleavage directions; conchoidal to splintery fracture	plaster of Paris, wallboard, soil conditioner
7.4–7.6	3 cleavage directions; irregular fracture	batteries, paints
5	no cleavage; conchoidal to splintery fracture	sulfuric acid

Appendix

Glossary

A

abrasion the grinding and wearing away of rock surfaces through the mechanical action of other rock or sand particles (33, 69)

acid precipitation rain, sleet, or snow that contains a high concentration of acids (35)

azimuthal projection (az uh MYOOTH uhl proh JEK shuhn) a map projection that is made by moving the surface features of the globe onto a plane (13)

B

beach an area of the shoreline made up of material deposited by waves (66)

bedrock the layer of rock beneath soil (42)

C

chemical weathering the process by which rocks break down as a result of chemical reactions (35)

conic projection a map projection that is made by moving the surface features of the globe onto a cone (12)

contour interval the difference in elevation between one contour line and the next (19)

contour line a line that connects points of equal elevation (18)

creep the slow downhill movement of weathered rock material (81)

cylindrical projection (suh LIN dri kuhl proh JEK shuhn) a map projection that is made by moving the surface features of the globe onto a cylinder (11)

D

deflation a form of wind erosion in which fine, dry soil particles are blown away (69)

differential weathering the process by which softer, less weather resistant rocks wear away and leave harder, more weather resistant rocks behind (39)

dune a mound of wind-deposited sand that keeps its shape even though it moves (70)

E

elevation the height of an object above sea level (18)

equator the imaginary circle halfway between the poles that divides the Earth into the Northern and Southern Hemispheres (7)

erosion the process by which wind, water, ice, or gravity transports soil and sediment from one location to another (49)

G

glacial drift the rock material carried and deposited by glaciers (76)

glacier a large mass of moving ice (72)

H

humus dark, organic material formed in soil from the decayed remains of plants and animals (44)

I

index contour on a map, a darker, heavier contour line that is usually every fifth line and that indicates a change in elevation (19)

L

landslide the sudden movement of rock and soil down a slope (79)

latitude the distance north or south from the equator; expressed in degrees (7)

leaching the removal of substances that can be dissolved from rock, ore, or layers of soil due to the passing of water (44)

loess (LOH ES) very fertile sediments of quartz, feldspar, hornblende, mica, and clay deposited by the wind (70)

longitude the distance east and west from the prime meridian; expressed in degrees (8)

M

magnetic declination the difference between the-magnetic north and the true north (6)

map a representation of the features of a physical body such as Earth (4)

mass movement a movement of a section of land down a slope (78)

mechanical weathering the breakdown of rock into smaller pieces by physical means (32)

mudflow the flow of a mass of mud or rock and soil mixed with a large amount of water (80)

P

parent rock a rock formation that is the source of soil (42)

prime meridian the meridian, or line of longitude, that is designated as 0° longitude (8)

R

relief the variations in elevation of a land surface (19)

remote sensing the process of gathering and analyzing information about an object without physically being in touch (15)

rock fall the rapid mass movement of rock down a steep slope or cliff (79)

S

saltation the movement of sand or other sediments by short jumps and bounces that is caused by wind or water (69)

shoreline the boundary between land and a body of water (63)

soil a loose mixture of rock fragments, organic material, water, and air that can support the growth of vegetation (42)

soil conservation a method to maintain the fertility of the soil by protecting the soil from erosion and nutrient loss (48)

soil structure the arrangement of soil particles (43)

soil texture the soil quality that is based on the proportions of soil particles (43)

stratified drift a glacial deposit that has been sorted and layered by the action of streams or meltwater (77)

T

till unsorted rock material that is deposited directly by a melting glacier (76)

topographic map (TAHP uh GRAF ik MAP) a map that shows the surface features of Earth (18)

true north the direction to the geographic North Pole (6)

W

weathering the process by which rock materials are broken down by the action of physical or chemical processes (32)

Spanish Glossary

A

abrasion/abrasión proceso por el cual las superficies de las rocas se muelen o desgastan por medio de la acción mecánica de otras rocas y partículas de arena (33, 69)

acid precipitation/precipitación ácida lluvia, aguanieve o nieve que contiene una alta concentración de ácidos (35)

azimuthal projection/proyección azimutal una proyección cartográfica que se hace al transferir las características de la superficie del globo a un plano (13)

B

beach/playa un área de la costa formada por materiales depositados por las olas (66)

bedrock/lecho de roca la capa de rocas que está debajo del suelo (42)

C

chemical weathering/desgaste químico el proceso por medio del cual las rocas se fragmentan como resultado de reacciones químicas (35)

conic projection/proyección cónica una proyección cartográfica que se hace al transferir las características de la superficie del globo a un cono (12)

contour interval/distancia entre las curvas de nivel la diferencia en elevación entre una curva de nivel y la siguiente (19)

contour line/curva de nivel una línea que une puntos que tienen la misma elevación (18)

creep/arrastre el movimiento lento y descendente de materiales rocosos desgastados (81)

cylindrical projection/proyección cilíndrica una proyección cartográfica que se hace al transferir las características de la superficie del globo a un cilindro (11)

D

deflation/deflación una forma de erosión del viento en la que se mueven partículas de suelo finas y secas (69)

differential weathering/desgaste diferencial el proceso por medio cual las rocas más suaves y menos resistentes al clima se desgastan y las rocas más duras y resistentes al clima permanecen (39)

dune/duna un montículo de arena depositada por el viendo que conserva su forma incluso cuando se mueve (70)

E

elevation/elevación la altura de un objeto sobre el nivel del mar (18)

equator/ecuador el círculo imaginario que se encuentra a la mitad entre los polos y divide a la Tierra en los hemisferios norte y sur (7)

erosion/erosión el proceso por medio del cual el viento, el agua, el hielo o la gravedad transporta tierra y sedimentos de un lugar a otro (49)

G

glacial drift/deriva glacial el material rocoso que es transportado y depositado por los glaciares (76)

glacier/glaciar una masa grande de hielo en movimiento (72)

H

humus/humus material orgánico obscuro que se forma en la tierra a partir de restos de plantas y animales en descomposición (44)

I

index contour/índice de las curvas de nivel en un mapa, la curva de nivel que es más gruesa y oscura, la cual normalmente se encuentra cada quinta línea e indica un cambio en la elevación (19)

L

landslide/derrumbamiento el movimiento súbito hacia abajo de rocas y suelo por una pendiente (79)

latitude/latitud la distancia hacia el norte o hacia el sur del ecuador; se expresa en grados (7)

leaching/lixiviación la remoción de substancias que pueden disolverse de rocas, menas o capas de suelo debido al paso del agua (44)

loess/loess sedimentos muy fértiles de cuarzo, feldespato, horneblenda, mica y arcilla depositados por el viento (70)

longitude/longitud la distancia hacia el este y hacia el oeste del primer meridiano; se expresa en grados (8)

M

magnetic declination/declinación magnética la diferencia entre el norte magnético y el norte verdadero (6)

map/mapa una representación de las características de un cuerpo físico, tal como la Tierra (4)

mass movement/movimiento masivo un movimiento hacia abajo de una sección de terreno por una pendiente (78)

mechanical weathering/desgaste mecánico el rompimiento de una roca en pedazos más pequeños mediante medios físicos (32)

mudflow/flujo de lodo el flujo de una masa de lodo o roca y suelo mezclados con una gran cantidad de agua (80)

P

parent rock/roca precursora una formación rocosa que es la fuente a partir de la cual se origina el suelo (42)

prime meridian/meridiano de Greenwich el meridiano, o línea de longitud, que se designa como longitud 0° (8)

R

relief/relieve las variaciones en elevación de una superficie de terreno (19)

remote sensing/teledetección el proceso de recopilar y analizar información acerca de un objeto sin estar en contacto físico con el objeto (15)

rock fall/desprendimiento de rocas el movimiento rápido y masivo de rocas por una pendiente empinada o un precipicio (79)

S

saltation/saltación el movimiento de la arena u otros sedimentos por medio de saltos pequeños y rebotes debido al viento o al agua (69)

shoreline/orilla el límite entre la tierra y una masa de agua (63)

soil/suelo una mezcla suelta de fragmentos de roca, material orgánico, agua y aire en la que puede crecer vegetación (42)

soil conservation/conservación del suelo un método para mantener la fertilidad del suelo protegiéndolo de la erosión y la pérdida de nutrientes (48)

soil structure/estructura del suelo la organización de las partículas del suelo (43)

soil texture/textura del suelo la cualidad del suelo que se basa en las proporciones de sus partículas (43)

stratified drift/deriva estratificada un depósito glacial que ha formado capas debido a la acción de los arroyos o de las aguas de ablación (77)

T

till/arcilla glaciárica material rocoso desordenado que deposita directamente un glaciar que se está derritiendo (76)

topographic map/mapa topográfico un mapa que muestra las características superficiales de la Tierra (18)

true north/norte verdadero la dirección al Polo Norte geográfico (6)

W

weathering/meteorización el proceso por el cual se desintegran los materiales que forman las rocas debido a la acción de procesos físicos o químicos (32)

Index

Index

Index

Index

Credits

PHOTOGRAPHY

Front Cover Brian Sytnyk/Masterfile

Skills Practice Lab Teens Sam Dudgeon/HRW

Connection to Astrology Corbis Images; **Connection to Biology** David M. Phillips/Visuals Unlimited; **Connection to Chemistry** Digital Image copyright © 2005 PhotoDisc; **Connection to Environment** Digital Image copyright © 2005 PhotoDisc; **Connection to Geology** Letraset Phototone; **Connection to Language Arts** Digital Image copyright © 2005 PhotoDisc; **Connection to Meteorology** Digital Image copyright © 2005 PhotoDisc; **Connection to Oceanography** © ICONOTEC; **Connection to Physics** Digital Image copyright © 2005 PhotoDisc

Table of Contents iv (cl), Tom Pantages Photography; iv (bc), Bob Krueger/Photo Researchers, Inc.; v (t), Sam Dudgeon/HRW; v (b), Tom Bean/CORBIS; vi–vii, Victoria Smith/HRW; x (bl), Sam Dudgeon/HRW; xi (tl), John Langford/HRW; xi (b), Sam Dudgeon/HRW; xii (tl), Victoria Smith/HRW; xii (bl), Stephanie Morris/HRW; xii (br), Sam Dudgeon/HRW; xiii (tl), Patti Murray/Animals, Animals; xiii (tr), Jana Birchum/HRW; xiii (b), Peter Van Steen/HRW

Chapter One 2–3, JPL/NASA; 4, Royal Geographical Society, London ,UK./The Bridgeman Art Library; 5 (t), Sam Dudgeon/HRW; 5 (b), Tom Pantages Photography; 6, Sam Dudgeon/HRW; 10 (bl, br), Andy Christiansen/HRW; 14, Texas Department of Transportaion; 15, Spaceimaging.com/Getty Images/NewsCom; 16 (bl, bc, br), Strategic Planning Office, City of Seattle; 16 (tl), HO/NewsCom; 17, Andy Christiansen/HRW; 18, USGS; 19 (tl), USGS; 19 (tr), USGS; 20, USGS; 23, Sam Dudgeon/HRW ; 25, USGS; 25 (br), Strategic Planning Office, City of Seattle; 28 (r), JPL/NASA; 28 (l), Victoria Smith/HRW; 29 (r), Bettman/CORBIS; 29 (bl), Layne Kennedy/CORBIS

Chapter Two 30–31, Johny Sundby/Zuma Press/NewsCom; 32, SuperStock; 33 (tc), Visuals Unlimited/Martin G. Miller; 33 (tl), Ron Niebrugge/Niebrugge Images; 33 (tr), Grant Heilman/Grant Heilman Photography; 34 (t), John Sohlden/Visuals Unlimited; 36 (t), Laurence Parent; 36 (b), C. Campbell/Westlight/Corbis; 37, Bob Krueger/Photo Researchers, Inc.; 38 (b), B. Ross/Westlight/Corbis; 40 (bl), Digital Image copyright © 2005 EyeWire ; 40 (br), David Cumming; Eye Ubiquitous/CORBIS; 41, Corbis Images; 42, The G.R. "Dick" Roberts Photo Library; 45, Tom Bean/Getty Images/Stone; 46 (t), Bill Ross/Westlight/Corbis; 46 (b), Bruce Coleman, Inc.; 47, Lee Rentz/Bruce Coleman, Inc.; 48 (bl), Grant Heilman Photography, Inc.; 48 (br), Charlton Photos, Inc.; 49, Kevin Fleming/CORBIS; 50 (tr), Mark Lewis/ImageState; 50 (tl), Paul Chesley/Getty Images/Stone; 50 (br), Tom Hovland/Grant Heilman Photography, Inc.; 50 (bl), AgStockUsa; 51, Bettmann/CORBIS; 52 (bl), 53 Sam Dudgeon/HRW; 54, B. Ross/Westlight/Corbis; 55, Bob Krueger/Photo Researchers, Inc.; 58 (tr), M.A. Kessler/Earth Sciences Department, University of California at Santa Cruz; 58 (tl), © W. Ming/UNEP/Peter Arnold, Inc.; 59 (t), Michael Murphy/By permission of Selah, Bamberger Ranch; 59 (b), Michael Murphy/By permission of Selah, Bamberger Ranch;

Chapter Three 61, John Kuntz/Reuters NewMedia Inc./CORBIS; 62, Aaron Chang/Corbis Stock Market; 63 (t), Tom Bean; 63 (b), CORBIS Images/HRW; 64 (tc), The G.R. "Dick" Roberts Photo Library; 64 (br), Jeff Foott/DRK Photo; 64 (bl), CORBIS Images/HRW; 65 (tl), Breck P. Kent; 65 (tr), John S. Shelton ; 66 (t), Don Herbert/Getty Images/Taxi; 66 (b), Jonathan Weston/ImageState; 66 (t), SuperStock; 67, InterNetwork Media/Getty Images; 69, Jonathan Blair/CORBIS; 70, Telegraph Colour Library/FPG International/Getty Images/Taxi; 72, Tom Bean/CORBIS; 74 (b), Getty Images/Stone; 74 (t), Visuals Unlimited/Glenn M. Oliver; 76, 77, Tom Bean; 78 (l), Sam Dudgeon/HRW; 78 (r), Sam Dudgeon/HRW Photo; 79 (b), Sebastian d'Souza/AFP/CORBIS; 79 (t), Jacques Jangoux/Getty Images/Stone; 80 (t), Jebb Harris/Orange County Register/SABA/CORBIS; 80 (b), Mike Yamashita/Woodfin Camp & Associates; 81, Visuals Unlimited/John D. Cunningham; 82, Sam Dudgeon/HRW; 84, Tom Bean/CORBIS; 85, Aaron Chang/Corbis Stock Market; 88 (bl), Geological Survey of Canada, Photo #2002–581, Photographer Dr. Rejean Couture; 88, Charles H. Stites/The Lost Squadron Museum; 88 (t), Louis Sapienza/The Lost Squadron Museum; 89 (r), ©National Geographic Image Collection/ Marla Stenzel; 89 (l), Martin Mejia/AP/Wide World Photos

Lab Book/Appendix "LabBook Header", "L", Corbis Images; "a", Letraset Phototone; "b", and "B", HRW; "o", and "k", images ©2006 PhotoDisc/HRW; 90, 91, Sam Dudgeon/HRW; 92, USGS; 93 – 96, Sam Dudgeon/HRW

TEACHER EDITION CREDITS

1E (t), Sam Dudgeon/HRW; 1E (b), HO/NewsCom; 1F (tl), Spaceimaging.com/Getty Images/NewsCom; 1F (bl), Strategic Planning Office, City of Seattle; 1F (r), USGS; 29E (l), SuperStock; 29E (tr), Laurence Parent; 29E (br), Corbis Images; 29F (l), The G.R. "Dick" Roberts Photo Library; 29F (r), Mark Lewis/ImageState; 59E (tl), Aaron Chang/Corbis Stock Market; 59E (bl), InterNetwork Media/Getty Images; 59E (r), Jonathan Blair/CORBIS; 59F (l), Tom Bean/CORBIS; 59F (r), Jebb Harris/Orange County Register/SABA/CORBIS

Answers to Concept Mapping Questions

The following pages contain sample answers to all of the concept mapping questions that appear in the Chapter Reviews. Because there is more than one way to do a concept map, your students' answers may vary.

CHAPTER 1 Maps as Models of the Earth

20.

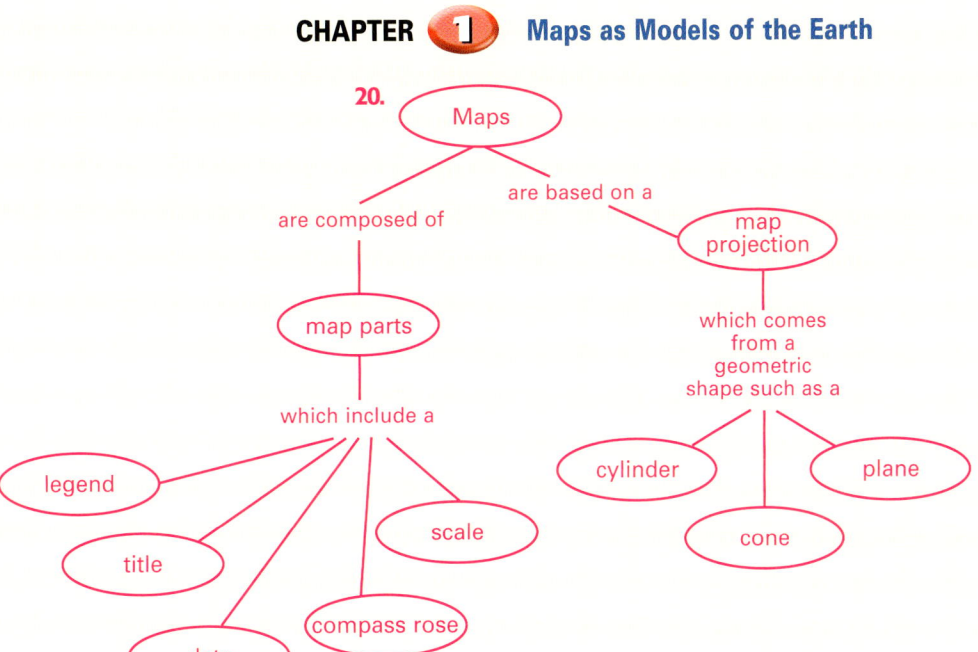

CHAPTER 2 Weathering and Soil Formation

17.

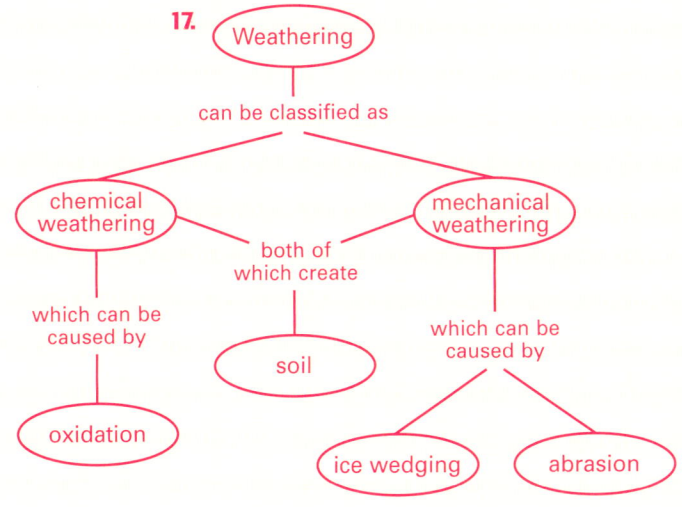

CHAPTER 3 Agents of Erosion and Deposition

20.